Advances in Intelligent Systems and Computing

Volume 992

The series "Advances in Intelligent Systems and Computing" contains publications on theory, applications, and design methods of Intelligent Systems and Intelligent Computing. Virtually all disciplines such as engineering, natural sciences, computer and information science, ICT, economics, business, e-commerce, environment, healthcare, life science are covered. The list of topics spans all the areas of modern intelligent systems and computing such as: computational intelligence, soft computing including neural networks, fuzzy systems, evolutionary computing and the fusion of these paradigms, social intelligence, ambient intelligence, computational neuroscience, artificial life, virtual worlds and society, cognitive science and systems, Perception and Vision, DNA and immune based systems, self-organizing and adaptive systems, e-Learning and teaching, human-centered and human-centric computing, recommender systems, intelligent control, robotics and mechatronics including human-machine teaming, knowledge-based paradigms, learning paradigms, machine ethics, intelligent data analysis, knowledge management, intelligent agents, intelligent decision making and support, intelligent network security, trust management, interactive entertainment, Web intelligence and multimedia.

The publications within "Advances in Intelligent Systems and Computing" are primarily proceedings of important conferences, symposia and congresses. They cover significant recent developments in the field, both of a foundational and applicable character. An important characteristic feature of the series is the short publication time and world-wide distribution. This permits a rapid and broad dissemination of research results.

**** Indexing: The books of this series are submitted to ISI Proceedings, EI-Compendex, DBLP, SCOPUS, Google Scholar and Springerlink ****

More information about this series at http://www.springer.com/series/11156

Mousumi Gupta · Debanjan Konar ·
Siddhartha Bhattacharyya ·
Sambhunath Biswas
Editors

Computer Vision and Machine Intelligence in Medical Image Analysis

International Symposium, ISCMM 2019

Springer

Editors
Mousumi Gupta
Department of Computer Applications
Sikkim Manipal Institute of Technology
East Sikkim, Sikkim, India

Debanjan Konar
Department of Computer Science
and Engineering
Sikkim Manipal Institute of Technology
East Sikkim, Sikkim, India

Siddhartha Bhattacharyya
Department of Information Technology
RCC Institute of Information Technology
Kolkata, West Bengal, India

Sambhunath Biswas
Machine Intelligence Unit
Indian Statistical Institute
Kolkata, West Bengal, India

ISSN 2194-5357 ISSN 2194-5365 (electronic)
Advances in Intelligent Systems and Computing
ISBN 978-981-13-8797-5 ISBN 978-981-13-8798-2 (eBook)
https://doi.org/10.1007/978-981-13-8798-2

This Springer imprint is published by the registered company Springer Nature Singapore Pte Ltd.
The registered company address is: 152 Beach Road, #21-01/04 Gateway East, Singapore 189721, Singapore

Dr. Mousumi Gupta *would like to dedicate the volume to her mother Smt. Ira Roy.*

Mr. Debanjan Konar *would like to dedicate the volume to his parents Mr. Debidas Konar and Mrs. Smritikana Konar who have encouraged him all the way and to his beloved daughter Sanvi.*

Dr. Siddhartha Bhattacharyya *would like to dedicate the volume to his father Late Ajit Kumar Bhattacharyya, his mother Late Hashi Bhattacharyya, his beloved wife Rashni Bhattacharyya and his colleagues Chandrayee, Amitabha, Prabhu, Rathin and Umesh.*

Preface

We feel greatly privileged to present the conference proceedings of the First International Symposium on Computer Vision and Machine Intelligence in Medical Image Analysis (ISCMM 2019) which was successfully held during February 26–27, 2019, in conjunction with the 2019 Second International Conference on Advanced Computational and Communication Paradigms (ICACCP) at Sikkim Manipal Institute of Technology, Sikkim, India.

Recently, visual computing-based analysis and inference have strengthened the medical field and are becoming an integral component to the community of medical practitioners. Embedded biomedical system performs a variety of task, starting from simple data acquisition and its display to various preprocessing tasks, such as noise rejection, and feature extraction followed by some complex decision-making processes based on sophisticated machine learning paradigms, helping out medical practitioners for effective treatment. Recent trends are dominantly found to be gravitated toward quantitative interpretation through computer methodologies. Algorithms, developed by scientists all over the globe, are gaining acceptance by the medical communities in the field of various disease diagnoses. However, as human lives are precious, we believe such medical image analysis algorithms should be carefully designed and tested and, therefore, need extra attention. Features should be very carefully chosen. It might so happen that benign diseases can mimic the behavior of malignant diseases or malignancies may sometimes exhibit characteristics of benign pattern, which could be misleading and fatal. Handling the possibilities is challenging. We, therefore, strongly believe that a common platform is badly needed in this area for large-scale discussion and interactions. The present conference is an attempt toward fulfilling this goal.

Scientists, engineers and clinicians exhibited keen interest and submitted papers in this conference on medical image processing. There were totally 55 papers submitted in the symposium from which 19 papers were accepted after strict scrutiny and out of which 15 papers were registered for oral presentation. The blind review process was initiated by assigning a minimum of three expert reviewers. The credit goes to the technical program committee for extending their support to arrange for the review process and making it complete within the notified time.

We also thank the local organizing committee for different conference-related activities. We believe without their active support the conference would not have been a success. Finally, we thank Springer publishing house for agreeing to publish the proceedings of our conference. We also thank the Medical Council of India and North Eastern Council for sponsoring the conference in various ways and, in particular, for their financial support in making the symposium possible.

Sikkim, India
February 2019

Mousumi Gupta
Debanjan Konar
Siddhartha Bhattacharyya
Sambhunath Biswas

Contents

About the Editors

Mousumi Gupta is Associate Professor in the Department of Computer Applications, Sikkim Manipal Institute of Technology (SMIT). She completed her M.Tech. and Ph.D. from SMIT in 2008 and 2015 respectively. Her area of research interest is digital image processing using pattern recognition. She has published research articles in international journals and conference proceedings. She is reviewer of six SCI & Scopus indexed journals and many conferences including IET Image processing and Journal for Electronic Imaging.

Debanjan Konar is Assistant Professor in the Department of Computer Science and Engineering, Sikkim Manipal Institute of Technology (SMIT). He completed his B.E. in Computer Science and Engineering from Burdwan University; M.Tech. in Multimedia and Software Systems from National Institute of Technical Teachers' Training & Research, Kolkata. Currently, he is pursuing Ph.D. from Indian Institute of Technology Delhi in the area of Quantum Inspired Soft Computing. He is reviewer of many journals like Applied soft computing, International Journal of Computers and Applications and International Journal of Hybrid Intelligence. He has been technical program committee member of many IEEE conferences.

Siddhartha Bhattacharyya [FIETE, SMIEEE, SMACM, SMIETI, LMCSI, LMOSI, LMISTE, MIET (UK), MIAENG, MIRSS, MIAASSE, MCSTA, MIDES, MIISIP] did his Bachelors in Physics, Bachelors in Optics and Optoelectronics and Masters in Optics and Optoelectronics from University of Calcutta, India in 1995, 1998 and 2000 respectively. He completed Ph.D. in Computer Science and Engineering from Jadavpur University, India in 2008. He is the recipient of the University Gold Medal from the University of Calcutta for his Masters. He is the recipient of the coveted Adarsh Vidya Saraswati Rashtriya Puraskar in 2016, Distinguished HoD and Distinguished Professor wards conferred by Computer Society of India, Mumbai Chapter, India in 2017 and Bhartiya Shiksha Ratan Award conferred by Economic Growth Foundation, New Delhi in 2017. He received the NACF-SCRA, India award for Best Faculty for Research in 2017. He received Honorary Doctorate Award (D. Litt.) from The University of South

America in 2017. He is also the recipient of the South East Asia Regional Computing Confederation (SEARCC) International Digital Award ICT Educator of the Year at Colombo, Sri Lanka in 2017. He is currently the Principal of RCC Institute of Information Technology, Kolkata, India. In addition, he is also serving as the Dean (Research and Development and Academic Affairs) of the institute. He is a co-author of 5 books and the co-editor of 30 books and has more than 220 research publications in international journals and conference proceedings to his credit. He has got a patent on intelligent colorimeter technology.

Sambhunath Biswas obtained an M.Sc. degree in Physics, and a Ph.D. in Radiophysics and Electronics, from the University of Calcutta in 1973 and 2001 respectively. In 1983–85 he completed courses of M.Tech. (Computer Science) from Indian Statistical Institute, Calcutta. His doctoral thesis is on Image Data Compression. He was a UNDP Fellow at MIT, USA to study machine vision in 1988–89 and was exposed to research in the Artificial Intelligence Laboratory. He visited the Australian National University at Canberra in 1995 and joined the school on wavelets theory and its applications. In 2001, he became a representative of the Government of India to visit China for holding talks about collaborative research projects in different Chinese Universities. He started his career in Electrical Industries, in the beginning as a graduate engineering trainee and, then as a design and development engineer. He was a System Analyst GR-I (in the rank of Professor) at the Machine Intelligence Unit in Indian Statistical Institute, Calcutta where he was engaged in research and teaching. He is at present the Professor and Head of the Computer Science & Engineering department of Techno India University, West Bengal, Calcutta. He is also an external examiner of the Department of Computer Science at Jadavpur University in Calcutta. He has published several research articles in International Journals of repute and is a member of IUPRAI (Indian Unit of Pattern Recognition and Artificial Intelligence). He is the principal author of a book titled Bezier and Splines in Image Processing and Machine Vision, published by Springer, London.

A Novel Method for Pneumonia Diagnosis from Chest X-Ray Images Using Deep Residual Learning with Separable Convolutional Networks

Rahul Sarkar, Animesh Hazra, Koulick Sadhu and Preetam Ghosh

Abstract Pneumonia is an infection that inflames the air sacs in one or both lungs. The air sacs may fill with fluid or pus, causing cough with phlegm or pus, fever, chills, and difficulty in breathing. Pneumonia can be caused by a variety of organisms such as bacteria, viruses, and fungi. Apart from causing difficulty in respiration, pneumonia can cause other complications such as bacteremia, lung abscess, and pleural effusion, among countless others. This paper presents a novel automated method for efficient and accurate pneumonia diagnosis from chest X-ray(CXR) images. Also, this model utilizes an efficient technique for noise reduction using bilateral filtering for edge preservation and optimum enhancement of the images using contrast-limited adaptive histogram equalization(CLAHE) to aid the detection of pneumonia clouds in the CXR images. The proposed model explores the benefits of deep residual learning along with separable convolution algorithm to achieve a classification accuracy of 98.82% and AUROC score of 0.99726 for diagnosing the disease. For cross-validation of the model, gradient-weighted class activation map(Grad-CAM) and saliency map visualization are used as a measure to verify the performance of the model along with localization of the affected regions in the CXR images.

Keywords Bilateral filtering · CLAHE · Global average pooling · Grad-CAM · Residual learning · Saliency mapping · Separable convolution · Vanishing gradient

---

R. Sarkar (✉) · A. Hazra · K. Sadhu · P. Ghosh
Jalpaiguri Government Engineering College, Jalpaiguri 735102, West Bengal, India
e-mail: rahulsarkar.2798@gmail.com

A. Hazra
e-mail: hazraanimesh53@gmail.com

K. Sadhu
e-mail: sadhu.koulickkumar@gmail.com

P. Ghosh
e-mail: preetam22n@gmail.com

M. Gupta et al. (eds.), *Computer Vision and Machine Intelligence in Medical Image Analysis*, Advances in Intelligent Systems and Computing 992,
https://doi.org/10.1007/978-981-13-8798-2_1

1 Introduction

Pneumonia remains the leading cause of death by infection among children under 5, accounting for about 2400 lives a day. Pneumonia alone accounted for about 16% of the 5.6 million deaths under the age of 5, killing around 880,000 children in 2016. Most of its victims were less than 2 years old [1]. Therefore, early detection of pneumonia is very important to avoid complications caused by the disease.

This project introduces a model which achieved an accuracy of 98.288% along with an AUC score of 0.9972 in diagnosing Pneumonia from CXR images. The proposed model can be found at https://github.com/RahulSkr/pneumoniaDetection.

The proposed methodology involves image denoising and enhancement using bilateral filtering and CLAHE algorithms, respectively, followed by resizing of the images as per the model input. Preprocessing of the images using the aforesaid algorithms with optimal parameters caused an increase in the validation accuracy of the model from 97.945 to 98.288%.

The chest X-ray dataset used here was available at Mendeley Data [2]. The model developed in this study utilizes the aforementioned dataset using the intuitions behind deep residual learning [3] alongside separable convolution algorithm [4]. The deep residual learning algorithm considerably reduces the problem of *vanishing gradient*, thereby avoiding stagnation in the learning phase of the model. Depth-wise separable convolution algorithm used in the proposed model greatly reduces the computation time as compared to the normal convolution algorithm.

In this study, validation of the proposed model's performance is done not only through the standard performance metrics but also its prediction ability is tested by visualizing the gradient weighted class activation map of the last convolution layer and the saliency map of the prediction layer. This method not only helps to validate the model, but also localizes the affected regions in the lungs to aid the diagnosis and following treatment.

The rest of the paper is organized as follows: Sect. 2 provides a brief overview of the literature survey that was performed prior to developing the model. Section 3 discusses the proposed methodology in details. Section 4 showcases the results and discussions to support the approach and finally Sect. 5 concludes the work along with its possible future scope.

2 Literature Survey

In this study, initially a survey work regarding the various techniques used for the model development was carried out, followed by a review of the previous works on pneumonia diagnosis using chest X-ray images. In this section, the summaries of some of the diagnosis techniques are presented in a nutshell.

Oliveira et al. [5] in their paper proposed a model named Pneumo-CAD, which uses a computer-aided diagnostic scheme over a wavelet transform. They present

Table 1 Summary of some of the existing statistical models used to detect pneumonia in CXR images

Authors	Year	Employed methodology	AUC score
Oliveira et al. [5]	2007	CAD and wavelet transform	0.9700
Sharma et al. [6]	2017	Otsu thresholding	–
Rajpurkar et al. [7][a]	2017	Deep learning	0.7680
Our methodology	–	Deep separable residual learning	0.9972

[a]Performed on the ChestX-ray14 dataset

two distinct methodologies in their work, both of which obtained best performances with Haar wavelet.

In 2017, Sharma et al. [6] used Otsu thresholding techniques as a means to detect Pneumonia clouds in the chest X-ray images. In the same year, a state-of-the-art model was introduced by Rajpurkar et al. [7], which utilized a deep learning model with a depth of 121 layer. This model was designed using the Chest X-ray14 dataset and is capable of accurately predicting probabilities of 14 diseases from a CXR image.

Kieu et al. [8] in 2018 introduced a deep learning approach to detect abnormalities in the CXR images. Bar et al. [9] introduced an approach of considerable importance in which a transfer learning model based on DeCAF is used for pathology detection in CXR images. A summary of some of the statistical pneumonia diagnosis techniques is provided in Table 1.

3 Proposed Methodology

Here, we introduce a novel deep learning algorithm for efficient and accurate diagnosis of pneumonia, and making it a state-of-the-art algorithm in the classification task under consideration on the given dataset. The proposed methodology has been illustrated in Fig. 1.

We then define the model which uses intuition behind the residual learning algorithm, to avoid the problem of vanishing gradient and utilizes the depth-wise separable convolution algorithm to reduce the parameter space and time complexity of the network.

3.1 Dataset Preparation

The dataset used in this study is available at Mendeley Data. This dataset consisted of 5856 chest X-ray images, where the images fall into one of the following three categories below.

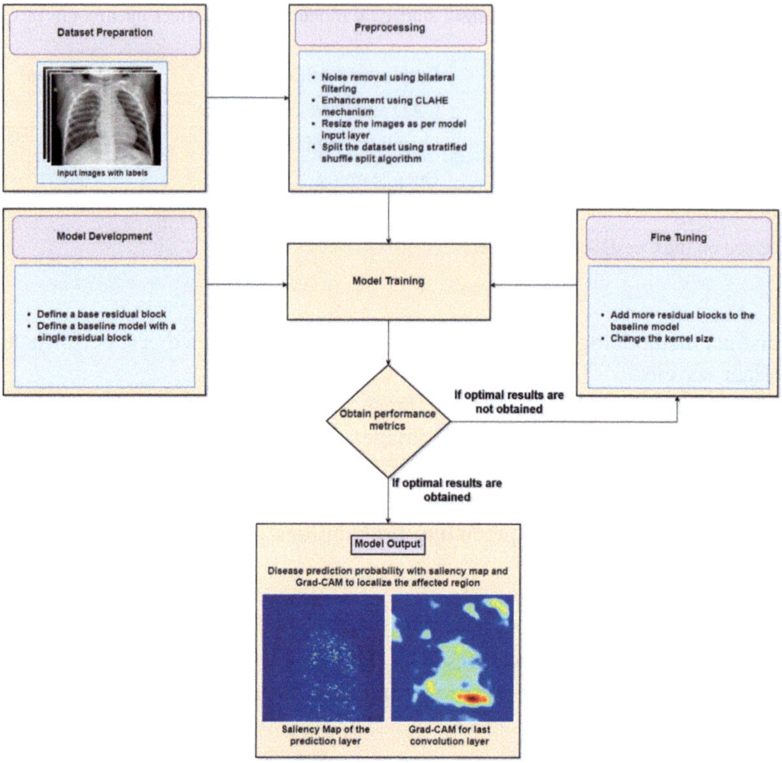

Fig. 1 Flowchart of the proposed methodology

- CXR images of patients not affected by pneumonia.
- CXR images of patients affected by bacterial pneumonia.
- CXR images of patients affected by viral pneumonia.

Visually distinguishing chest X-ray images of patients affected by bacterial pneumonia from those affected by viral pneumonia is not advised, and can only be diagnosed accurately via cough analysis. Hence, we develop a binary classifier which can accurately diagnose the presence of pneumonia clouds in CXR images.

3.2 *Data Preprocessing*

Data preprocessing is essential for analyzing the medical image data. Proper noise removal and enhancement techniques applied to raw images can result in significant improvement in the performance of the model. Here, we feed the data to the model after removing noise from the images and then enhancing the images using the algorithms with best-observed performance. Finally, the images were resized to fit the model optimally.

3.2.1 Noise Removal from the Images

Noise removal is a crucial part of image preprocessing on image data to improve the results. Bilateral filtering was used in order to denoise the images, which performs efficient smoothening of the images while preserving the edges. Bilateral filter [10, 11] is defined as follows:

$$BF(I, x) = \frac{1}{W_p} \sum_{x_i \epsilon \Omega} I(x_i) f_r(\|I(x_i) - I(x)\|) g_s(\|x_i - x\|) \tag{1}$$

Here, W_p is the normalization term defined in the following way:

$$W_p = \sum_{x_i \epsilon \Omega} f_r(\|I(x_i) - I(x)\|) g_s(\|x_i - x\|) \tag{2}$$

In Eq. 2, I represents a k-dimensional image, Ω is the set of pixels x_i in a n^k window, f_r is the range attenuation function and g_s is the spatial attenuation function.

We found the best parameters for denoising the images producing the least amount of intermediate levels of gray values around the edges which is one of the complications that arise by applying the bilateral filtering as addressed by C. Tomasi.

3.2.2 Enhancement of the Images

After noise removal from the images, image enhancement was performed using CLAHE mechanism. Wen et al. [12] explains how in the CLAHE mechanism each pixel in the image performs an equalization on its histograms via a rectangular window enclosing it, the brightness of the image is then redistributed and hence causes a change in the contrast of the image. Further, we refer to the results of CLAHE enhancement on CXR images as proposed by Koonsanit et al. [13]. CLAHE mechanism is hence preferred because it avoids over-enhancement of the contrast and amplification of noise in the images.

3.2.3 Resizing the Images

Finally, the images are resized to an optimal size to avoid excess information loss from the images. Here, the optimal results are obtained for the dimension (224, 224, 3). The images are read in BGR format and contrast stretching was performed by normalizing the pixel intensity values. The normalization equation used here is defined

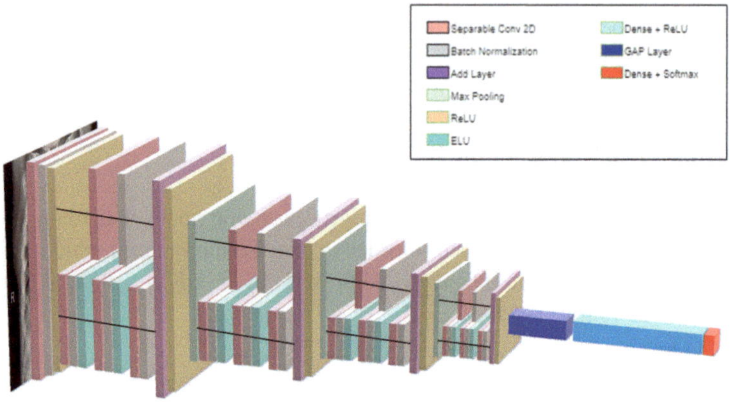

Fig. 2 A 3D view of the proposed model

below.

$$I_N = (I - Min)\frac{newMax - newMin}{Max - Min} + newMin \tag{3}$$

Here, (Min, Max) are the existing limits for intensity range of I and $(newMin, newMax)$ are the new limits for the new intensity range of I_N. In this study, the images are normalized to have intensity values within the range of 0–1.

3.3 *Model Development*

A 3D view of the proposed model is provided in Fig. 2. It is seen from the view that the model consists of 4 distinct blocks where the concept of residual learning is used. Each of these blocks contains an identity mapping which is essential for efficient propagation of the error through the network, hence avoiding vanishing gradient problem. All the convolution layers in the model follow the depth-wise separable convolution algorithm.

Here, the GAP layer is used rather than a *flatten* layer to obtain a 1D array from the convolution feature maps, this highly reduces the parameter complexity of the model. However, it comes at the cost of losing some information from the feature maps. Here, we use max-pooling layers throughout the model to minimize the size of the feature maps of the ultimate convolution layer.

3.3.1 Designing the Residual Block

From Fig. 2, it is clear that the residual branch consist of two activated convolution layers followed by an inactive convolution layer. It is this layer to which a identity shortcut connection is mapped. This mapping does not add any extra parameters to the model and cause no change in its runtime. This shortcut connection ensures an end-to-end training of the network, and hence aids error minimization.

The forward propagation through the activation function which follows the "Add layers" (say l) in the proposed network is defined in as follows:

$$a^l := g(W^{l-1,l} \cdot a^{l-1} + b^l + W^{l-3,l} \cdot a^{l-3}) \tag{4}$$

Here, a^l is the activation of units in the layer l, g(ReLU in this case) is the activation function of layer l and $W^{l-k,l}$ is the weight matrix between layers $l-k$ and l. Here, we add $W^{l-3,l}$ to the activation function of layer l in the residual block because the shortcut path maps the $l-3$ layer onto the l layer, this reduces the degradation of accuracy with the increase in depth.

The backward propagation of error from layer $l-k$ to l can be generally defined as

$$\Delta w^{l-k,l} := -\eta a^l \cdot \delta^l \tag{5}$$

Here, η is the learning rate, a^l is the activation and δ^l is the loss signal of units in the layer l. Considering our model, $\Delta w^{l-1,l}$ from the residual path is degraded more compared to $\Delta w^{l-3,l}$ via the *skip path*(this problem of degradation of the gradient of error function is known as vanishing gradient problem). Hence, existence of the *skip path* or identity mapping ensures effective backpropagation through the network.

3.3.2 Separable Convolution Algorithm Used in the Network

Separable convolution algorithm requires $O(w \times d)$ runtime and parameter space whereas a traditional convolution algorithm requires $O(w^d)$, where w is the kernel width and d is the dimension of the input feature space. The convolution operation is defined as

$$y[m,n] = h[m,n] * x[m,n] = \sum_{j=-\infty}^{\infty} \sum_{i=-\infty}^{\infty} h[i,j] \cdot x[m-i, n-j] \tag{6}$$

If separable convolution is applied then

$$h[m,n] = h_1[m] \cdot h_2[n] \tag{7}$$

Substituting Eq. 7 in Eq. 6, we obtain

$$y[m,n] = \sum_{j=-\infty}^{\infty} h_2[j] \Big(\sum_{i=-\infty}^{\infty} h_1[i] \cdot x[m-i, n-j] \Big) \tag{8}$$

Here, $x[m, n]$ is the input signal and $h[m, n]$ is the impulse response. From Eq. 8, it is clear that separable convolution algorithm in 2D space is essentially an 1D convolution performed in horizontal and vertical directions. Thus, separable convolution algorithm simplifies and boosts the convolution operation as it requires more kernels for computation.

3.4 Visualization and Fine Tuning

In this study, we find the best parameters for the model using grid search and further verify the model's performance using Grad-CAM to visualize the last convolution layer, as the outputs of this layer are fed forward to the fully connected layers for prediction. Saliency maps were obtained for the prediction layer, which helps in visualizing the pneumonia clouds in the lung air sacs.

4 Result and Discussion

Model validation is crucial to prove the effectiveness of a model. For the proposed model, we not only use the standard performance metrics but also use certain visualization techniques to improve the model's quality. The three main performance measures which are considered in this study are as follows:

- Confusion matrix
- Receiver operating characteristics(ROC) curve
- Model visualization via Grad-CAM and saliency map

The above performance measures have been discussed in the sections below in detail.

4.1 Confusion Matrix

Plotting the confusion matrix for a model is an efficient way of visualizing the model's performance. The confusion matrix shown in Fig. 3 is generated on the test set, which comprises of 1168 samples from the original dataset containing 5856 images.

The performance of the proposed model against a corresponding plain network with traditional convolution algorithm is shown in Table 2.

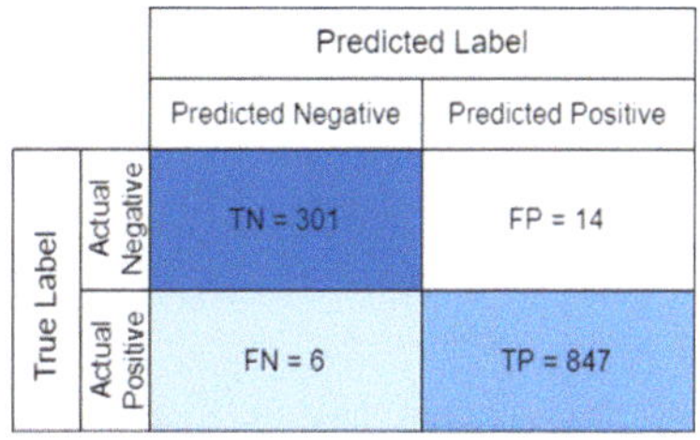

Fig. 3 Plot of the un-normalized confusion matrix for the proposed model

Table 2 Comparison of performances of the proposed model and a plain model

Performance metrics of the model	Formula of performance metrics	Plain convolution model for (3 * 3) kernel size	Proposed model for (3 * 3) kernel size
Accuracy(ACC)	$ACC = \frac{TP+TN}{TP+TN+FP+FN}$	0.97431	0.98288
Precision(P)	$P = \frac{TP}{TP+FP}$	0.98242	0.98374
Sensitivity(R)	$R = \frac{TN}{TP+FN}$	0.98242	0.99297
Specificity(S)	$S = \frac{TN}{TN+FP}$	0.95238	0.95555
F1 score(F_1)	$F_1 = \frac{2\times TP}{2\times TP+FP+FN}$	0.98241	0.98833

4.2 *Performance Curves*

Performance curves of the proposed model are illustrated in Fig. 4 along with the performance curves of the plain model for reference.

It becomes clear from Fig. 4a, b, that the proposed model proves to be considerably more accurate than its plain counterpart. In case of the plain network, the loss saturates at a considerably higher value giving an inaccurate result unlike the proposed network. This occurs because of the problem of vanishing gradient as already discussed in Sect. 3.3.1. Additionally, the proposed model using separable convolution took approximately **180** s per epoch during training, whereas the plain model which uses traditional convolution took **252** s because separable convolution utilizes more kernels in comparison and is hence faster.

Figure 4c shows comparison of the ROC curves of the proposed model and the corresponding plain model. It is clear that the proposed model has achieved a near ideal state, with an AUC score of 0.9972 whereas its plain counterpart achieved a score of 0.9952.

An optimal precision–recall trade-off was also obtained by the proposed model against the plain model which is shown in Fig. 4d.

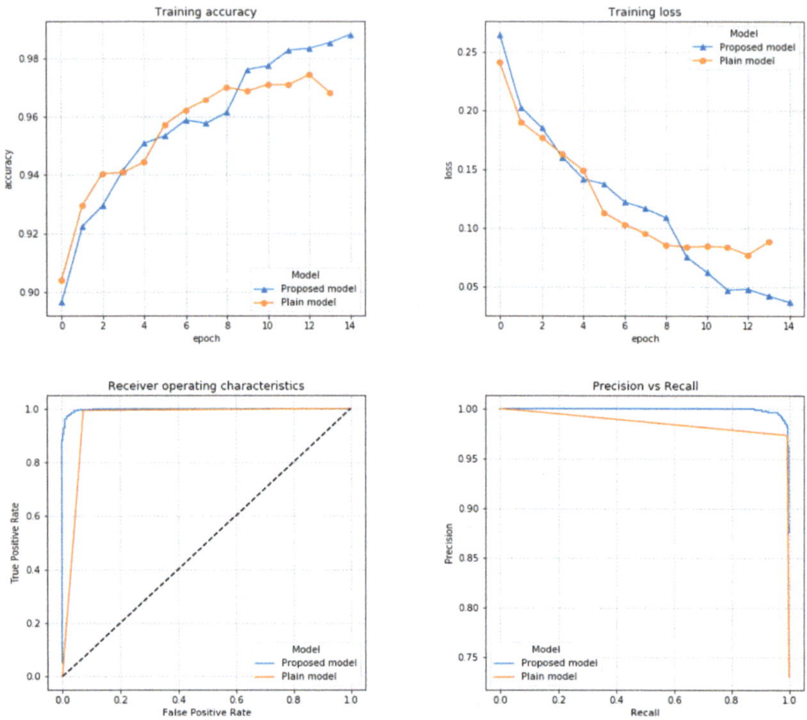

Fig. 4 **a** and **b** show the comparison of training accuracy and training loss curves of the proposed and plain model respectively, **c** comparison of ROC curves of the proposed and plain model, **d** comparison of precision–recall curves of the proposed and plain model

4.3 *Grad-CAM and Saliency Map Visualization*

Gradient weighted class activation map or simply Grad-CAM takes the gradients of a class that is passed to the final convolution layer and produces a localization map that highlights the important regions for predicting that class.

Saliency maps, on the other hand, highlight the pixels in the input, a small change in which can cause a change in the output prediction of the model. The Grad-CAMs and saliency maps of a few chest X-ray images are shown in Fig. 5.

5 Conclusion and Future Scope

In 2012, out of the 28,952 deaths from pneumonia in the UK, 58 were aged under 14 years, 1,374 were among those aged between 15–64 and 27,520 were among those aged 65 and above. Early diagnosis of pneumonia is hence of paramount importance. In this study, we devise an algorithm which significantly reduces parameter cost and

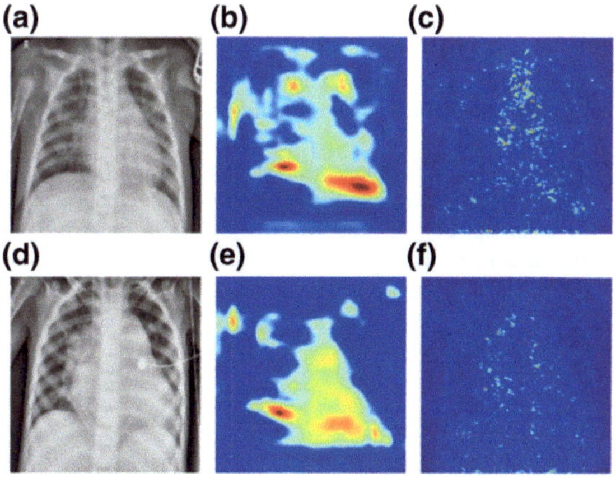

Fig. 5 (**a**)–(**b**): Processed images, (**c**)–(**d**): Grad-CAMs of the processed images, (**e**)–(**f**): Saliency maps of the processed images

minimizes the model error. The proposed method utilizes the deep residual learning algorithm and hence has a cross-entropy loss lower than and an execution time higher than its plain counterpart. The accuracy of the plain network reached a maximum of 97.431% as opposed to our proposed model which achieved an accuracy of 98.288%. The traditional convolution algorithm resulted in a computation time, which is 1.1526 times higher than our proposed method that uses separable convolution algorithm.

In future, we are planning to utilize our algorithm and develop a multi-label, multi-class classification model to evaluate the performance of the said algorithm on the benchmark dataset ChestXray14.

References

1. UNICEF: Pneumonia, UNICEF Data, June 2018. https://data.unicef.org/topic/child-health/pneumonia/
2. Kermany, D., Zhang, K., Goldbaum, M.: Labeled optical coherence tomography (OCT) and chest X-Ray images for classification, Mendeley Data, vol. 2 (2018). https://doi.org/10.17632/rscbjbr9sj.2
3. He, K., Zhang, X., Ren, S., Sun, J.: Deep residual learning for image recognition. arXiv:1512.03385
4. Chollet, F.: Xception: deep Learning with Depthwise Separable Convolutions, https://doi.org/10.1109/CVPR.2017
5. Oliveira, L.L.G., et al.: Computer-aided diagnosis in chest radiography for detection of childhood pneumonia. Int. J. Med. Inform. **77**, 555–564 (2008)
6. Sharma, A., Raju, D., Ranjan, S.: Detection of pneumonia clouds in chest X-ray using Image processing approach. In: Nirma University International Conference on Engineering (2017)
7. Rajpurkar, P., et al.: CheXNet: radiologist-level pneumonia detection on chest x-rays with deep learning. arXiv:1711.05225
8. Kieu, P.N., et al.: Applying multi-CNNs model for detecting abnormal problem on chest x-ray images. In: International Conference on Knowledge and Systems Engineering, pp. 300–305
9. Bar, Y., et al.: Chest pathology detection using deep learning with non-medical training. In: International Symposium on Biomedical Imaging (ISBI), pp. 294–297

10. Tomasi, C., Manduchi, R.: Bilateral filtering for gray and color images. In: Sixth International Conference on Computer Vision, pp. 839–846 (1998)
11. Banterle, F., Corsini, M., Cignoni, P., Scopigno, R.: A low-memory, straightforward and fast bilateral filter through subsampling in spatial domain. Comput. Graph. Forum **31**(1), 19–32 (2012)
12. Wen, H., Qi, W., Shuang, L.: Medical X-ray image enhancement based on wavelet domain homomorphic filtering and CLAHE. In: International Conference on Robots & Intelligent System, pp. 249–254 (2016)
13. Koonsanit, K., Thongvigitmanee, S., Pongnapang, N., Thajchayapong, P.: Image enhancement on digital x-ray images using N-CLAHE. https://doi.org/10.1109/BMEiCON.2017.8229130

Identification of Neural Correlates of Face Recognition Using Machine Learning Approach

Shreya Gupta and Tapan Gandhi

Abstract Object recognition has always been one of the key areas of research in modern times, especially in healthcare and engineering industry, because of the wide range of applications it has. There have been various methods to recognize and classify objects like shape matching, color matching, sliding window approach, etc. but a common problem the computational models face is the appropriate representation of 3D objects, image variation with angle variation, illumination effects, and the high computational costs of these models to maintain output accuracy. In this paper, we propose a novel method to detect and analyze the process of object recognition from magnetoencephalogram (MEG) signals of human brain. This could be made possible by classifying the object as face/scrambled face using machine learning through support vector machine (SVM). We train our SVM model using the recordings in DecMeg Human Brain dataset obtained from Kaggle. In addition, by calculating the accuracy of individual sensors for the duration, we are able to identify the cluster of sensors responsible for visual recognition and the dynamic interaction among sensors with the passage of time using neural coordinates of the magnetometer sensors with an accuracy of 74.85%. We found that the sensors in the occipitotemporal and occipitoparietal lobes are most actively involved in visual classification. The proposed approach has been able to reduce the previous effective time stamp of 100–360 ms to 124–240 ms. This reduces the computational cost of the model while establishing the essential relationships between MEG signals and facial detection.

S. Gupta (✉)
Delhi Technological University, New Delhi, Delhi, India
e-mail: shreyagupta_bt2k16@dtu.ac.in

T. Gandhi
Indian Institute of Technology, New Delhi, Delhi, India
e-mail: tgandhi@iitd.ac.in

M. Gupta et al. (eds.), *Computer Vision and Machine Intelligence in Medical Image Analysis*, Advances in Intelligent Systems and Computing 992,
https://doi.org/10.1007/978-981-13-8798-2_2

1 Introduction

Face recognition is one of the most critical skills that human beings develop from the very beginning of our childhood. Humans are very good at recognizing faces in challenging viewing conditions like complex background, across large image transformation and in various illuminating conditions. Face recognition, hence, plays an important role in healthcare and engineering applications. In engineering, the application was first formulated for the purpose of criminal identification [1]. Its use and application have since exponentially increased, and have spread to fields like retail [2], auto photo tagging [3], robot navigation [4], video surveillance [5], and marketing [6]. In health care, face recognition plays a very significant role in diagnosis of neurodevelopmental disorders like autism, prosopagnosia, and delusional misidentification syndromes such as Capgras syndrome. These vulnerable groups have a hampered social interaction because of their inability to recognize faces [7–9].

Several methods and algorithms have been proposed to perform human face recognition, for example [10, 11] use artificial neural network, [12] uses the concept of eigenfaces to recognize new faces in an unsupervised manner and [13] applies mosaic grid approach, among others. Previous researchers have majorly worked on identification and recognition tasks [1–5]. However, Yang and Huang [13] have used knowledge-based recognition approach for face location in complex images, using a mosaic grid approach; Turk [12] uses face images projected in a feature space that encodes the variation among known face images and defines them by a set of eigenvectors.

A challenge that the existing methodologies have faced so far include poor performance of the algorithms on large datasets, inability to deal with illumination effects, and low tolerance to image variations [14]. Since human beings are able to identify objects irrespective of lighting and angle variations, recent times have, thus, seen an increase in devising algorithms which are closer to the functioning of human brain object recognition system. This paper adopts and extends the approach by finding and building a relationship between neurons in human brain and visual recognition of objects, in particular face. The work attempts to do so using support vector machine's (SVM) property of maximizing the distance between classes.

Section 2 describes the experimental setup and data preprocessing technique used. Section 3 presents the selection criteria (of the sensors) and the results obtained. Section 4 discusses the future scope and opens doors for further research in this domain.

2 Methodology

In this section, we elucidate on the algorithm chosen to fit the data, the experimental setup, and processing performed before training and testing the model.

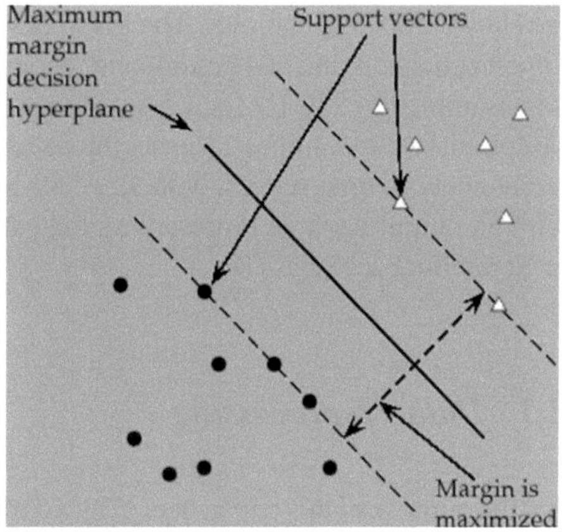

Fig. 1 Hyperplane separating binary classes [12]

2.1 Support Vector Machines

SVM is a supervised machine learning algorithm. It is believed to be the best "off-the-shelf" algorithm, especially for high-dimensional spaces. SVM uses decision boundary, also called as separating hyperplane, to distinguish between various classes of the target variable. The hypothesis function (h) for an SVM is given by

$$h_{w,b}(x) = g(w^T x + b) \tag{1}$$

where w and b are hyperparameters described in detail in [15]; $g(z) = 1$ if $z >= 0$, and $g(z) = -1$ otherwise.

This distinction can be linear, Gaussian, or polynomial depending upon the type, dimensions, and complexity of dataset. The data points closest to the decision boundary are called support vectors. SVM maximizes the distance between these support vectors and the separating hyperplane obtained before, using the most appropriate kernel (Fig. 1). Due to space constraints, we refer the readers to [15] for more technical details on SVMs and kernels. This paper performs a binary classification into face/scrambled face and after thorough analysis, Gaussian (or RBF) kernel given by the following equation is used:

$$K(x, z) = exp(-(\ ||x - z||^2\ )\ /\ (2 \times \sigma^2)) \tag{2}$$

2.2 Experimental Setup

The dataset chosen is "DecMeg Humain Brain" dataset from publicly available Kaggle repository [16]. The dataset contains the magnetoencephalography (MEG) recordings of 23 subjects (16 train subjects and 7 test subjects) when each of them are shown 580–590 samples. The MEG device records 306 time series (102 magnetometer channels and 204 gradiometer channels), one for each channel at a frequency downsampled to 250 Hz for 1.5 s (starting 0.5 s before the stimulus starts) for each sample shown to a subject. Hence, the dataset is a three-dimensional matrix (sample $\times$ channels $\times$ time series), with size $580 \times 306 \times 275$. Each sample is associated with an output variable representing the category of visual stimulus, Face (class 1) or Scramble Face (class 0).

2.3 Data Preprocessing

For the purpose of this paper, data of all 16 train subjects is taken. The work calculates the mean of MEG recordings obtained during the initial 0.5 s time stamp and subtracts the mean from the entire data to normalize it. Post extracting data from the actual time of stimulus (last 1 s), a new normalized input data is created.

For each subject, the dataset is divided into batches of 20 samples, with a 50% overlap and 7:3 hold out cross validation. After standardizing the dataset and using a box constant, $c = 0.65$ on a RBF kernel, SVM is trained and test set accuracy and false alarm results are saved in a 102×16 matrix (magnetometer sensors $\times$ subjects). Results are obtained by taking average accuracy of each sensor and color mapping these accuracies in a two-dimensional scalp plot of each subject, for each time slot.

3 Results

The resultant accuracy scalp plots are shown in Fig. 2. The accuracy obtained on testing the MEG recordings of individual sensors as a response to visual stimulus for a designated time slot is indicative of two things:

1. Relevance of the sensor in face recognition.
2. Distinction with which the sensor detected an object in designated time period, i.e., the relevance of time slot in face recognition.

3.1 Sensor Selection Criteria

MEG readings (in fT) for the sample labeled as face and as scrambled face against time (in ms) are plotted in Fig. 3a, b, respectively. As is evident from these figures, it is not possible to distinguish and typecast the two signals or the effective time periods. The accuracy obtained using Gaussian kernel finds the most active sensors and their most responsive time slots. The work uses the benchmark accuracy of 55%

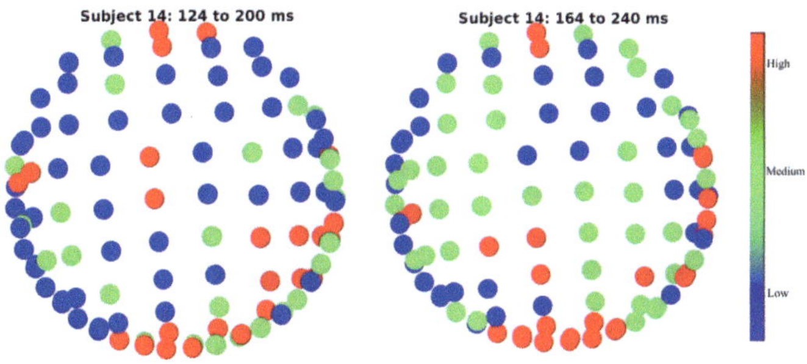

Fig. 2 2D scalp plot during 124–200 ms and 164–240 ms time stamp

Fig. 3 MEG readings for sample detected as face and scrambled face

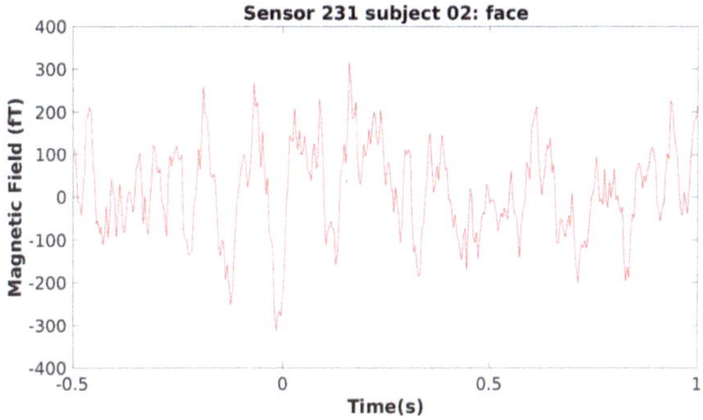

(a) face

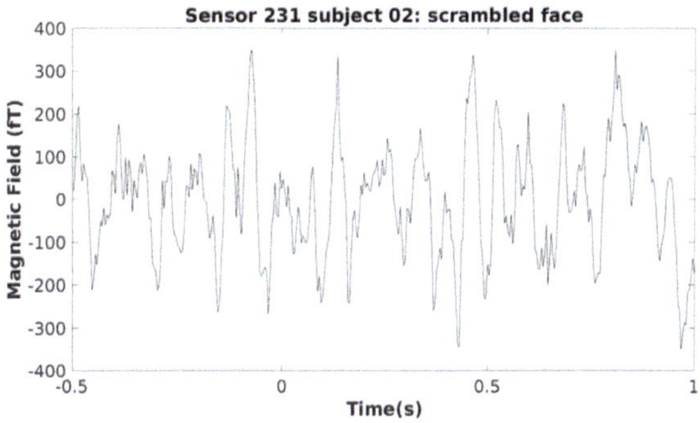

(b) scrambled face

to classify the sensor as an active sensor (marked in red). Sensors having accuracy between 52 and 55% are less participating sensors (marked in green) and those with accuracy lesser than 52% are poor sensors (marked in blue). Figures 2 and 4 are some of the plottings obtained when these sensors and their coordinates, obtained from source [20], are color mapped with their accuracy.

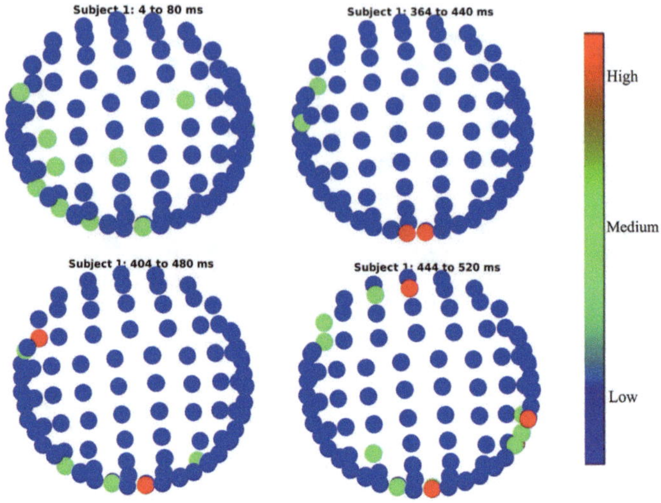

Fig. 4 Scalp plot during less responsive time stamps

3.2 *Results and Discussion*

As can be seen in Fig. 2, most sensors marked red (i.e., having higher significance) are located near occipitotemporal and occipitoparietal lobes and few in frontal lobe. To compare the results, the scalp plot for other (less responsive) time slots is shown in Fig. 4 and accuracy achieved in a time stamp is plotted in Fig. 5. As can be seen from Fig. 5, the maximum accuracy increases significantly during 120–240 ms.

Thus, in the process of identifying the most receptive sensors and the time slot of their highest accuracy, the work has been able to identify and model, using the proposed SVM approach, the neural correlates of facial recognition. The model, thus generated, was able to detect and distinguish face and scrambled face with an accuracy of 74.85%. This accuracy was obtained by sensors in the occipitotemporal and occipitoparietal lobes of a human brain during 124–240 ms time stamp. The relatively low accuracy is sensical since it is obtained by interpolating results from individual sensors and as the common knowledge persists, the brain decodes an object using a cluster of such sensors.

By identifying the active sensors and the differentiating time stamps, the present work has been able to filter out noisy and less effective sensors and time slots. Comparing the results with previous benchmarks [19] (Table 1), we can see the proposed work reduces the effective time stamp from 1000 ms to 120 ms (i.e., by 88%) as compared to 260 ms achieved in [19]. This mitigates the computational costs of the model built for face recognition (using principles of neuroscience) by providing researchers the channels and slots to focus their research on.

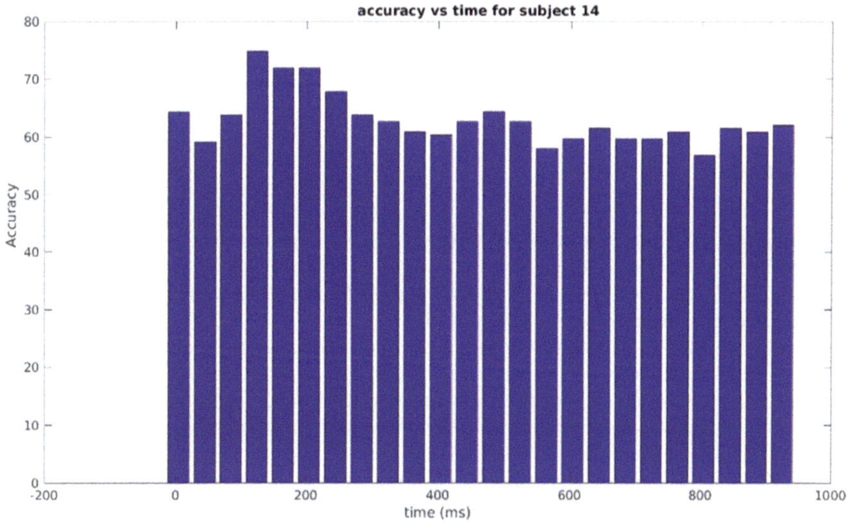

Fig. 5 Accuracy of the sensors in different time slots

Table 1 Comparison of effective time stamps

Approach	Active time stamps (in ms)	% Trimmed
Marcus et. al. [19]	100–360	74
Proposed work	120–240	88

4 Conclusion and Future Scope

Using neuroscience for the process of face recognition, as shown, might have given less accuracy at the moment but its results are independent of illumination and angle effects which are the major drawbacks of traditional face recognition approaches. Consistent with the past studies, the work is able to find, using the proposed machine learning model, the time stamps in which brain is able to distinguish between a special class of object, i.e., face. Hence, the trade-off between computationally less expensive baseline model and higher accuracy makes way for heavier models to be executed above it. It can also be inferred that human brain is able to differentiate objects based on certain features. The findings of this paper can be extended to extract features captured by the eyes during the mentioned time stamps (124–240 ms) which enabled brain to detect and classify faces. The central aim is to provide a mechanism that can artificially assist people suffering from deficit of face recognition and thereby to restore their social interaction. The feature identification can be performed using a multimodal approach like eye tracking while visualizing brain signal recording simultaneously and can then be used to design a more sophisticated and optimum model using AI and computer vision.

Acknowledgements We would like to acknowledge Kaggle for providing us with human brain signal dataset for facilitation of the research proposed in this paper.

References

1. Burton, A.M., Wilson, S., Cowan, M., Bruce, V.: Face recognition in poor-quality video: evidence from security surveillance. Psychol. Sci. **10**, 243–248 (1999)
2. Wu, C.-C., Zeng, Y.-C., Shih, M.-J.: Enhancing retailer marketing with an facial recognition integrated recommender system. IEEE (2015)
3. Stone, Z., Zickler, T., Darrell, T.: Autotagging facebook: social network context improves photo annotation. In: Workshop on Internet Vision (2008)
4. Boucenna, S., Gaussier, P., Andry, P., Hafemeister, L.: A robot learns the facial expressions recognition and face/non-face discrimination through an imitation game. Int. J. Soc. Robot. (2014)
5. Lei, Z., Wang, C., Wang, Q., Huang, Y.: Real-time face detection and recognition for video surveillance applications. In: 2009 World Congress on Computer Science and Information Engineering, pp. 168–172. (2009)
6. Shergill, G.H., Sarrafzadeh, H., Diegel, O., Shekar, A.: Computerized sales assistants: the application of computer technology to measure consumer interest;a conceptual framework. J. Electron. Commer. Res. **9**(2), 176191 (2008)
7. Klin, A., Sparrow, S.S. De Bildt, A., Cicchetti, D.V., Cohen, D.J., Volkmar, F.R.: A normed study of face recognition in autism and related disorders. J. Autism Dev. Disord. 499–508 (1999)
8. Grter, T., Grter, M., Carbon, C.C.: Neural and genetic foundations of face recognition and prosopagnosia. J. Neuropsychol. 79–97 (2008)
9. Pierce, K., Mller, R.-A., Ambrose, J., Allen, G., Courchesne, E.: Face processing occurs outside the fusiform 'face area' in autism: evidence from functional MRI. Brain **124**(10), 2059–2073 (2001)
10. Bruce, V., Young, A.: Understanding face recognition. Br. J. Psychol. **77**(3), 305–327 (1986)
11. Baron, R.J.: Mechanisms of human facial recognition, Int . J. Man-Mach. Stud. **15**, 137–178 (1981)
12. Le, T.H.: Applying artificial neural networks for face recognition. In: Advances in Artificial Neural Systems (2011)
13. Turk, M., Pentland, A.: Face recognition using eigenfaces. In: Proceedings of the IEEE Computer Society Conference on Computer Vision and Pattern Recognition, pp. 586–591 (1991)
14. Yang, G., Huang, T.S.: Human face detection in a complex background. Pattern Recognit. **27**, 53–63 (1994)
15. Khurana, P., Sharma, A., Singh, SN., Singh, P.K.: A survey on object recognition and segmentation techniques. In: 3rd International Conference on Computing for Sustainable Global Development (INDIACom), pp. 3822–3826 (2016)
16. CS229: Machine Learning, retrieved from chapter notes on Support Vector Machines. http://cs229.stanford.edu/syllabus.html
17. Manning, C., Raghavan, P., Schtze, H.: Introduction to Information Retrieval. Cambridge University Press, Chapter 15 (2008)
18. DecMeg2014: Decoding the Humain Brain. https://www.kaggle.com/c/decoding-the-human-brain/data
19. https://github.com/FBKNILab/DecMeg2014
20. Streit, M., Dammers, J., Simsek-Kraues, S., Brinkmeyer, J., Wolwer, W., Ioannides, A.: Time course of regional brain activations during facial emotion recognition in humans. Neurosci. Lett. **342**(12), 101–104 (2003)

An Overview of Remote Photoplethysmography Methods for Vital Sign Monitoring

Ruchika Sinhal, Kavita Singh and M. M. Raghuwanshi

Abstract Vital signs such as heart rate, respiratory rate, blood pressure, body temperature, and oxygen saturation are essential for early detection of any related significant illness. Many of the existing methods that are used to monitor the aforementioned vital signs are camera-based. In these methods, sensors are fixed to the body that are not sturdy to the motion of the subject. Another method to monitor vital signs is photoplethysmography (PPG), an emerging noncontact technique that maps, spatial blood volume variation in living tissue from the images captured through a video. Most of the camera-based methods are driven by three remote photoplethysmography algorithms. The camera-based methods are useful for detecting vital signs with an objective AAA, i.e., anyone, anywhere, and anytime. However, there exist few challenges in r-ppg methods and make it an open research problem. This paper presents an overview of the signal processing challenges faced by remote photoplethysmography for calculating the vital signs with a focus on heart rate estimation.

Keywords Heart rate · Remote photoplethysmography · Vital signs · Signal processing · Video processing

---

R. Sinhal (✉)
Department of CSE, Datta Meghe Institute of Engineering, Technology & Research, Wardha, India
e-mail: ruchisinhal04@gmail.com

K. Singh
Department of Computer Technology, Yeshwantrao Chavan College of Engineering, Nagpur, India
e-mail: singhkavita19@yahoo.co.in

M. M. Raghuwanshi
Department of Information Technology, Yeshwantrao Chavan College of Engineering, Nagpur, India
e-mail: m_raghuwanshi@rediffmail.com

M. Gupta et al. (eds.), *Computer Vision and Machine Intelligence in Medical Image Analysis*, Advances in Intelligent Systems and Computing 992,
https://doi.org/10.1007/978-981-13-8798-2_3

1 Introduction

The human body has many integral signs such as heart rate, respiratory rate, blood pressure, body temperature, and oxygen saturation, to name a few that helps in deciding the fitness of the human body. Measurement of these vital signs needs a device, and therefore, it is not possible to measure them for the examination of body fitness at home. The heart rate is one of the important vital signs, as it plays a vital role in checking the fitness level of a human body. Monitoring of the heart rate even helps in detecting any growing heart problem. Heart rate is measured as beats per minute (bpm) that ranges from 60 to 100 bpm in healthy adults.

Respiratory rate (RR) is another vital sign, defined as the frequency of breaths per minute which ranges from 12 to 20 breaths per minute in a normal person. Monitoring respiration can give valuable information about neural and pulmonary conditions. Checking human RR regularly is an important task of examining the health of a person [1]. Measuring these mentioned vital signs needs some medical devices and a doctor to examine them. Generally, an individual has to visit the clinic and record the signs. For instance, medical professionals examine heart rate through ECG, which is the most commonly used contact-based or invasive method that induces skin irritation.

The noninvasive healthcare monitoring system involves either capturing the image or recording the video through a digital camera. One of the systems is photoplethysmography (PPG) that is used to detect blood volume variations in the microvascular tissue. PPG devices have been competent to monitor relevant signs that include pulse rate, respiration rate, and body temperature [2]. Verkruysse et al., in [3], described image photoplethysmography (i-PPG) technique. In this paper, the authors recorded the facial skin affected by the port-wine dye after laser therapy. The resulting data were used to obtain maps of the amplitude and phase of the spatially differing PPG signals. The main drawback of the proposed technique is that the PPG signals are sensitive to variations induced due to motion when camera-based phones are used. Many a time, the noninvasive vital sign monitoring is also termed as remote photoplethysmography (r-PPG). The name r-PPG has been coined due to the monitoring of remotely captured video through digital camera. Therefore, in the forthcoming paragraph, we present an overview of the work presented by various authors related to PPG vital sign monitoring.

Independent component analysis (ICA), a blind source separation method proposed by Poh et al., in [4] removed the noise from PPG signal face imaging. The measuring standards are recommended for the use of ECG sensors for measuring the heart rate variability (HRV) in [5]. However, it has been shown that PPG-derived heart rate variability can be a good substitute for HRV at rest [6]. In [7, 8], Sun et al. compared the performance of a low-cost web camera and a high-sensitivity camera to evaluate the variability of the heart rate and pulses. The authors stated that the 30 fps webcam function is similar to the 100 fps camera when signals are incorporated to improve the time-domain resolution [8]. HRV has been used to detect real-time changes in the workload to evaluate and index the autonomous nervous system [9]. Its spectral analysis can provide a sympathetic balance, a ratio that reflects mutual changes in sympathetic and vagal outflows [10]. HRV tends to be rhythmic and emotionally positive, followed by a phenomenon known as respiratory sinus arrhythmia. HRV, on the other hand, tends to chaotic, angry, anxiety, or sadness. These rhythmic variations create a condition of cardiac coherence [11, 12]. Detection of physiological signals using noncontact equipment is especially beneficial in emotional computing, where emotions like stress or fear are induced. Contact sensors can create a bias in these physiological experiments by interfering with the user, which results in an erroneous emotion classification [13]. Currently, published methods effectively recover spectral components of HR, BR, and HRV over a certain period. However, there have been a few attempts, instant HR measurement with a webcam, particularly considering artifacts of head motion [14].

The r-PPG algorithms proposed in the literature have been developed on videos under constrained environments. However, there are many challenging issues faced during the development of algorithms under uncontrolled environment. Therefore, it is a potential field of research for researchers who are willing to work in the field of r-PPG for vital sign monitoring. These challenges have been described in Sect. 3.

This paper organization has three sections. The introductory section describes the vital signs and its importance. Section 2 discusses various r-PPG methods used for estimating the heart rate. Section 3 describes the different challenges faced in r-PPG. Finally, concluding section presents a summary of the review of techniques.

2 r-PPG Methods for Estimating Heart Rate

r-PPG is a remote photoplethysmography technique that measures, small changes in skin color caused by variations, in volume and oxygen saturation while heart pumping. All r-PPG techniques developed so far have used the videos captured from a digital camera for analyzing the pulse or heart rate. Recently, several r-PPG algorithms are developed for extracting the heart signal from videos. In this section, we present the study of each approach developed by researchers for estimating heart rate. Broad categories of the various approaches are blind source separation, model-based methods, and design-based methods. Each category and algorithm is further discussed in the forthcoming section.

2.1 Blind Source Separation (BSS)

The generalization of time series data as an alternative representation in the frequency domain is also important. This representation enables the understanding of the signals and the filtering or interpolation of the data. In particular, the singular value decomposition (SVD) [4] and independent component analysis (ICA) [15] techniques for the principal component analysis (PCA) have been examined. Both these PCA and ICA techniques use statistical data representation rather than time or frequency domain. In other words, data are projected on a new set of axes that fulfill certain statistical criteria, which implies independence, instead of a set of axes representing discrete frequencies such as the Fourier transformation, where independence is assumed. The criterion depends on the structure of the data and the axes on which the data is projected. The projection direction increases the signal-to-noise ratio, which allows us to observe the important structural signals. For example, the power spectrum of the data can be calculated to make the peaks of certain signals visible and to separate the noise from the signal. Such unwanted signals can be filtered using PCA and ICA. Most important, BSS techniques are analytical and computational for general problems of signal processing. It does not benefit from the unique characteristic of skin reflections used to solve the r-PPG problem. The ICA-based approach, in particular, normalizes the standard deviation of RGB signals, ignoring the fact that the PPG signal induces distant yet known relative amplitudes in the particular RGB channels.

2.2 *Model-Based Method*

The BSS method discussed above has limitations of assumption on the colors associated with source signals. In the blind source separation method, the colors are considered independently for the signal estimation. To overcome the limitations of BSS, model-based methods use different components of color vectors to control de-mixing. Therefore, model-based methods eliminate the dependency of colors on skin color reflection including light color. In addition, the model-based methods are also motion tolerance. Model-based method includes PBV and CHROM techniques proposed by De Haan in [15, 16], respectively. The PBV technique is based on blood volume pulse that retrieves the pulse directly from the pulsatile components restricting all color variations to possible direction. The PBV is also a motion robust improved method which uses blood volume pulse signature as mentioned. Further, the CHROM technique is robust to motion based on the standardized assumption of skin tone. The CHROM is different from PBV because it reduces sensitivity by eliminating the specular component and reducing the size. The CHROM algorithm assumes a standardized skin tone vector that allows white images to be balanced. From the literature, it has been observed that the CHROM is robust to mono-white illuminations and so it is categorized as best model-based algorithm. In addition, Wenjing Wang proposed a new method, the orthogonal skin plane (POS) [17]. This method resembles CHROM but alters the order in which the expected color distortions are reduced using different priors. In this new algorithm, authors developed a skin tone orthogonal plane in a temporarily normalized RGB environment.

Compared to multistep CHROM and POS, PBV is a one-step process and requires an accurate knowledge of the signature of the blood volume pulse. With regard to movement and stationary parameters, CHROM and POS perform well in stationary and motion situations when the alpha tuning is driven either by pulse or by large distortions, while PBV is specifically designed for movement. In addition, CHROM and POS are not as restrictive as PBV. In addition to all the above comparative analysis, one more similarity between CHROM and POS is that these two methods use soft priors to define a projection plane for alpha tuning in blood volume pulsation (i.e., channel ranking).

2.3 Design-Based Method

Model-based methods are good as they are robust to motion. They perform better than non-model-based methods. As discussed earlier, the limitation in CHROM is that it uses the vector for white skin reference, whereas PBV depends on the blood signals and diverts the signal to that side. However, apart from these limitations, the vital sign measurement is best for model-based methods as these methods are motion robust.

In the recently developed spatial subspace rotation (2SR) method [18], the RGB values are quantified as spatial representation. In the temporal domain, pulsatile blood causes variation in RGB channels, thus changing subspace of skin pixels. The algorithm creates a subject-dependent skin color space and tracks the tone change over time to measure the pulse in which the instant tone is determined on the basis of the statistical distribution of the skin pixels in the image. The idea of using the hue as a basic pulse extraction parameter is supported by the analysis of the use of different color spaces to measure the pulse [19]. Since then, the tone drives measurement, the method at an early stage eliminates all variations in intensity. In this sense, 2SR is a skin approach that defines a temporarily normalized orthogonal projection plane in the RGB pulse extraction space. The subspace axes built by 2SR are, however, completely data-driven without physiological consideration. This presents performance problems in practice when spatial measurements are not reliable, i.e., when the skin mask is noisy or poorly selected. A new lock-in technique is proposed in [20] for extracting pulse rate which when compared with gold standards differed only by four beats [20].

In this section, we presented different approaches proposed by various researchers in last few decades. From the detailed literature, it has been observed that estimation of heart rate using r-PPG from captured video still struggles with many challenges. These challenges are further discussed in the next forthcoming section.

3 Challenges in r-PPG

From the literature, it is evident that r-PPG focuses on extracting the pulse signal from video to estimate the heart rate of a person. However, r-PPG has many challenges due to various factors like subject motion, ambient light illumination, image optimization, spectrum analysis, etc. These factors present challenges to the researcher for recovering the accurate physiological data for different r-PPG methods. The men-

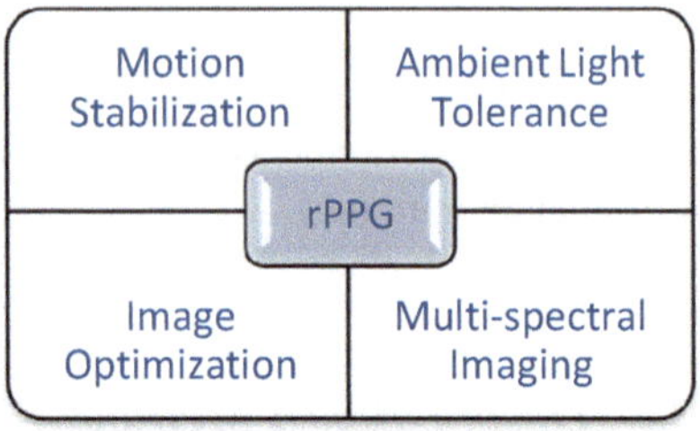

Fig. 1 Factors related to challenges in r-PPG

tioned four factors related to the main challenges in r-PPG are shown in Fig. 1. In subsequent sections, we present each of these factors for better understanding.

3.1 Motion Stabilization

The subject motion has been studied in most of the r-PPG algorithms to analyze its effects. The subject motion is an important and a great challenge that is faced by r-PPG algorithms. The subject motion changes the region of interest in the video when the subject is in motion. Early investigations of r-PPG focused on rigid and stationary regions of interest as used in researches [3, 14, 16, 21–23]. Ideally, when subjects are stationary, the ROI must also be constant in subsequent frames used by r-PPG algorithms. However, this might not be the practical scenario of the application of r-PPG algorithms. Therefore, there is a need of motion stabilization that is to be achieved in the video. Thus, the tracking of ROI can be proposed as a solution to this problem.

On the other hand, few of the studies introduced later focus on face imaging of the subjects that allowed limited naturalistic motion [19, 21]. The work proposed in [19, 21] used the emotion factor in unconditional environment while recording the videos for the study.

Further studies investigated the algorithm's performance under translational motion which limited the use of techniques as simple region of interest (ROI). These ROIs focused on object tracking [15], color difference, and chrominance-based signals from RGB color space [19]. Later, approaches for estimating the motion artifacts and correcting the obtained r-PPG signal using adaptive filtering have also been explored [24]. As stated earlier, this motion artifact is a challenge for r-PPG, with specific advancements in three big areas: (1) the development of an algorithm for image processing, (2) spatial redundancy, and (3) the use of integrated multiband techniques in visible and invisible wavelengths. While it is important to explore systematically varying motion artifacts, use cases in applied environments are included but not limited to exercise [25, 26], cognitive states [27, 28], and clinical care [29, 30], with a view to transition from the laboratory environment. Finally, the effects of movement artifacts have been extended beyond the pulse rate alone to cardiopulmonary measures. The authors in [31] proposed a new technology sub-band r-PPG for HR measurement in fitness by increasing the pulse extraction signal dimensionality. During fitness exercise, they tested the algorithm on the subjects. The proposed pre-filtering method improved r-PPG performance. The approach of selective amplitude filtering filtered the r-PPG signal based on the RGB color band. The authors in [32, 33] designed a filtering method that filters the RGB signals before the pulse is extracted.

3.2 Ambient Light Tolerance

Ambient light is the light already present in a video without any manipulation. There is no additional light added in the video. Such ambient lighting conditions might be considered proportionately consistent in most of the applications; nevertheless, there are some probable use cases like a computer simulation, virtual reality, mobile screen brightness, etc, where there can be variations in lighting conditions. The difference in illumination intensity many a time influences the intensity of the PPG waveform [23]. At the same time, the effects of ambient light on any other currently available PPG methods are unknown. In a limited, uncontrolled study in [32], Li et al. used an adaptive filtering approach, with more background region of interest. The ROI served as the input noise reference signal which compensates for background illumination. The results in [34] depicted improvement for varying illumination on a publicly available video database [35]. The detailed experiments presented in [34] derived modest reduction in heart rate error when compared to ECG recorded.

To estimate the vital signs, it is very important to deal with the ambient light tolerance in the background of the video. This is because the background or the region of interest, if it is either under-illuminated or over-illuminated, might result in incorrect signal extraction. This incorrect signal extraction will further lead to an incorrect estimate from the video. Consequently, it is very important to keep the illumination of a region of interest constant throughout the video acquisition process. The background illumination might be canceled using the technique developed in [34] as one of the illumination cancelation techniques. However, identification and removal of the effect of illumination on the region of interest still remains a challenging issue to be explored that further bring complexity in analyzing the heart rate from the region of interest.

3.3 Image Optimization

Image optimization specifically emphasizes on capturing the images or videos through different sources. The varieties of image sensors or video cameras that allow fostering of r-PPG also bring variations in a different level. For instance, a variation in the image sensor (e.g., digital camera, mobile phone cameras, etc,) brings variations in features that are to be examined, which additionally induces variations in further analyses. These features could be a basic sensor type, color separation sensors, special sensors, aspect ratio, image sizes, and the number of pixels, to name a few.

However, from the literature, it can be seen that the image quality does not rely only on the quality of an image sensor being used. The other factors that affect the image quality are a type of lens (CCD and CMOS), spectral properties as an illumination source, and image shutter speed. Thus, image quality directly or indirectly changes the feature of an image or video under consideration. Apart from this, although the frame rate is not directly related to image sensor properties, it is also considered as one of the vital components for variations in features under consideration for estimating vital signs. An image can be optimized with the use of automatic ROI selection from the image. Recently, Wang et al. performed supervised living skin detection using r-PPG by transforming the video into signal shape descriptor called multiresolution iterative spectrum [36].

Thus, an image optimization deals with capturing video in different formats. One such format for video is 'mp4' format that stores the frames in compressed form. The compression of an image or a video might lose information resulting in an inaccurate vital sign estimation from the recorded video. In this context, the image optimization is, therefore, a challenge and is to be dealt with before measuring the vital sign from the optimized video.

3.4 *Multispectral Imaging*

Multispectral imaging captures images with a specific wavelength and multiband spectrum. There are different spectral bands used for satellite images, such as Blue, Green, Red, near-infrared, mid-infrared, and thermal. Many contemporary r-PPG studies focus on three spectral bands in the visible light spectrum, i.e., red, green, and blue. While green/orange visible bands are the most pervasive of common oxygen and de-oxyhemoglobin derivatives [37], multispectral imagery from a single image sensor has often been used for r-PPG methods involving linear decomposition using multiple data channels. As shown by Martinez et al., in [38] with front spectrophotometry, some wavebands are better for measuring r-PPG pulse rate and respiration rate. The different spectrums have a different effect on the signal estimation. Thus, the r-PPG is still an open challenge because the different bands lead to the creation of different estimates for vital signs. This shows that it is difficult to identify good signal for estimation of vital signs.

4 Conclusion

Photoplethysmography technique maps the blood volume pulsating signals into the vital signs. The vital signs can be measured by heart rate, respiratory rate, or saturation per oxygen level (SPO2) from images or videos. In this paper, we have presented three r-PPG methods, namely, BSS, model-based, and design-based, based on digital video signals. Each method has its own advantages and limitations in various contexts of r-PPG as discussed in the paper. Among all the three described methods, BSS is a widely used r-PPG method in the literature. Further, we presented different factors, like subject motion, ambient light illumination, image optimization, and spectrum analysis, that make r-PPG an open research issue to be explored. Among all the factors listed above, motion stabilization is an important factor to be dealt with as it brings an uncontrolled environment and makes r-PPG a challenging task.

From the discussions presented in this paper, it has been observed that many research challenges are still open for solutions in the field of r-PPG. The researchers can initiate to pursue research and contribute to social technology.

References

1. Sinhal, R., Singh, K., Shankar, A.: Estimating vital signs through non-contact video-based approaches: a survey. In: 2017 International Conference on Recent Innovations in Signal processing and Embedded Systems (RISE), pp. 139–141. IEEE, Bhopal (2017)
2. Allen, J.: Photoplethysmography and its application in clinical physiological measurement. Physiol. Meas. **28**, R1–R39 (2007)
3. Verkruysse, W., Svaasand, L.O., Nelson, J.S.: Remote plethysmographic imaging using ambient light. Opt. Express **16**, 21434 (2008)
4. Poh, M.-Z., McDuff, D.J., Picard, R.W.: Advancements in noncontact, multiparameter physiological measurements using a webcam. IEEE Trans. Biomed. Eng. **58**, 7–11 (2011)
5. Camm, A.J., et al.: Heart rate variability. Standards of measurement, physiological interpretation, and clinical use. Task force of the european society of cardiology and the north american society of pacing and electrophysiology. Eur. Heart J. **17**, 354–381 (1996)
6. Gil, E., Orini, M., Bailón, R., Vergara, J.M., Mainardi, L., Laguna, P.: Photoplethysmography pulse rate variability as a surrogate measurement of heart rate variability during non-stationary conditions. Physiol. Meas. **31**, 1271–1290 (2010)
7. Sun, Y., Papin, C., Azorin-Peris, V., Kalawsky, R., Greenwald, S., Hu, S.: Use of ambient light in remote photoplethysmographic systems: comparison between a high-performance camera and a low-cost webcam. J. Biomed. Opt. **17**, 037005 (2012)
8. Sun, Y., Hu, S., Azorin-Peris, V., Kalawsky, R., Greenwald, S.: Noncontact imaging photoplethysmography to effectively access pulse rate variability. J. Biomed. Opt. **18**, 061205 (2012)
9. Appelhans, B.M., Luecken, L.J.: Heart rate variability as an index of regulated emotional responding. Rev. Gen. Psychol. **10**, 229–240 (2006)
10. Hoover, A., Singh, A., Fishel-Brown, S., Muth, E.: Real-time detection of workload changes using heart rate variability. Biomed. Signal Process. Control **7**, 333–341 (2012)
11. Servant, D., Logier, R., Mouster, Y., Goudemand, M.: Heart rate variability. Appl. Psychiatry. L'Enceph. **35**, 423–428 (2009)
12. Tiller, W.A., McCraty, R., Atkinson, M.: Cardiac coherence: a new, noninvasive measure of autonomic nervous system order. Altern. Ther. Health Med. **2**, 52–65 (1996)
13. Pavlidis, I., Dowdall, J., Sun, N., Puri, C., Fei, J., Garbey, M.: Interacting with human physiology. Comput. Vis. Image Underst. **108**, 150–170 (2007)
14. Bousefsaf, F., Maaoui, C., Pruski, A.: Continuous wavelet filtering on webcam photoplethysmographic signals to remotely assess the instantaneous heart rate. Biomed. Signal Process. Control **8**, 568–574 (2013)

15. de Haan, G., Jeanne, V.: Robust pulse rate from chrominance-based rPPG. IEEE Trans. Biomed. Eng. **60**, 2878–2886 (2013)
16. de Haan, G., van Leest, A.: Improved motion robustness of remote-PPG by using the blood volume pulse signature. Physiol. Meas. **35**, 1913–1926 (2014)
17. Wang, W., den Brinker, A., Stuijk, S., de Haan, G.: Algorithmic principles of remote-PPG. IEEE Trans. Biomed. Eng. 1–1 (2016)
18. Wang, W., Stuijk, S., de Haan, G.: A novel algorithm for remote photoplethysmography: spatial subspace rotation. IEEE Trans. Biomed. Eng. **63**, 1974–1984 (2016)
19. Tsouri, G.R., Li, Z.: On the benefits of alternative color spaces for noncontact heart rate measurements using standard red-green-blue cameras. J. Biomed. Opt. **20**, 048002 (2015)
20. Eaton, A., Vishwanath, K., Cheng, C.-H., Paige Lloyd, E., Hugenberg, K.: Lock-in technique for extraction of pulse rates and associated confidence levels from video. Appl. Opt. **57**, 4360 (2018)
21. Humphreys, K., Ward, T., Markham, C.: Noncontact simultaneous dual wavelength photoplethysmography: a further step toward noncontact pulse oximetry. Rev. Sci. Instrum. **78**, 044304 (2007)
22. McDuff, D.J., Estepp, J.R., Piasecki, A.M., Blackford, E.B.: A survey of remote optical photoplethysmographic imaging methods. In: 2015 37th Annual International Conference of the IEEE Engineering in Medicine and Biology Society (EMBC), pp. 6398–6404. IEEE, Milan (2015)
23. Shi, P., Hu, S., Zhu, Y.: A preliminary attempt to understand compatibility of photoplethysmographic pulse rate variability with electrocardiogramic heart rate variability. (2008)
24. Wieringa, F.P., Mastik, F., van der Steen, A.F.W.: Contactless multiple wavelength photoplethysmographic imaging: a first step toward "SpO2 camera" technology. Ann. Biomed. Eng. **33**, 1034–1041 (2005)
25. Sun, Y.: Motion-compensated noncontact imaging photoplethysmography to monitor cardiorespiratory status during exercise. J. Biomed. Opt. **16**, 077010 (2011)
26. Takano, C., Ohta, Y.: Heart rate measurement based on a time-lapse image. Med. Eng. Phys. **29**, 853–857 (2007)
27. McDuff, D., Gontarek, S., Picard, R.: Remote measurement of cognitive stress via heart rate variability (2014)
28. Cennini, G., Arguel, J., Akşit, K., van Leest, A.: Heart rate monitoring via remote photoplethysmography with motion artifacts reduction. Opt. Express **18**, 4867 (2010)
29. Aarts, L.A.M., Jeanne, V., Cleary, J.P., Lieber, C., Nelson, J.S., Bambang Oetomo, S., Verkruysse, W.: Non-contact heart rate monitoring utilizing camera photoplethysmography in the neonatal intensive care unit—A pilot study. Early Hum. Dev. **89**, 943–948 (2013)
30. Klaessens, J.H., van den Born, M., van der Veen, A., Sikkens-van de Kraats, J., van den Dungen, F.A., Verdaasdonk, R.M.: Development of a baby friendly non-contact method for measuring vital signs: First results of clinical measurements in an open incubator at a neonatal intensive care unit (2014)
31. Wang, W., den Brinker, A.C., Stuijk, S., de Haan, G.: Robust heart rate from fitness videos. Physiol. Meas. **38**, 1023–1044 (2017)
32. Wang, W., den Brinker, A.C., Stuijk, S., de Haan, G.: Color-distortion filtering for remote photoplethysmography (2017)
33. Wang, W., den Brinker, A.C., Stuijk, S., de Haan, G.: Amplitude-selective filtering for remote-PPG. Biomed. Opt. Express. **8**, 1965 (2017)
34. Li, X., Chen, J., Zhao, G., Pietikainen, M.: Remote heart rate measurement from face videos under realistic situations (2014)
35. Soleymani, M., Lichtenauer, J., Pun, T., Pantic, M.: A multimodal database for affect recognition and implicit tagging. IEEE Trans. Affect. Comput. **3**, 42–55 (2012)
36. Wang, W., Stuijk, S., de Haan, G.: Living-skin classification via remote-PPG. IEEE Trans. Biomed. Eng. **64**, 2781–2792 (2017)
37. Zijlstra, W.G., Buursma, A., Meeuwsen-van der Roest, W.P.: Absorption spectra of human fetal and adult oxyhemoglobin, de-oxyhemoglobin, carboxyhemoglobin, and methemoglobin. Clin. Chem. **37**, 1633–1638 (1991)

38. Corral Martinez, L.F., Paez, G., Strojnik, M.: Optimal wavelength selection for noncontact reflection photoplethysmography (2011)

Fuzzy Inference System for Efficient Lung Cancer Detection

Laxmikant Tiwari, Rohit Raja, Vaibhav Sharma and Rohit Miri

Abstract This paper suggests a lung cancer detection system with marked and unmarked nodules of cancerous elements that are detected and classified. Identification of lung cancer in earlier stage will reduce the cause of death, and lung cancer is said to be one of the leading causes of death. Computed tomography (CT) is used for lung cancer analysis and diagnosis, and manual process suffers from several challenges such as poor accuracy. There are numerous research contributions in this area but research attempt toward robustness is all-time challenge. We have implemented a fuzzy inference system, which includes four important stages as preprocessing, image segmentation, feature extraction, and design of fuzzy inference rules. These rules are used to identify cancerous cells accurately.

Keywords Lung nodule · Fuzzy inference system · Image segmentation

1 Introduction and Background

Numerous literature suggest that lung cancer has become a common problem, a major challenge, and one of the leading causes of death, especially in developing countries. The diagnosis that involves detection of lung nodule exactly is the area of focus for the researchers carrying out research and development activities. Usually, small lesions cannot be identified with the help of X-rays, and therefore computed tomography (CT) is used for the detection of lung cancer. Large number of cases, less

L. Tiwari
Department of CSIT, Dr. C. V, RAMAN University, Bilaspur, India

R. Raja (✉)
Department of CSE, Sreyas Institute of Engineering and Technology, Nagole, Hyderabad, India
e-mail: rohit.raja@sreyas.ac.in

V. Sharma
Department of Information Technology, Dr. C. V, RAMAN University, Bilaspur, India

R. Miri
Department of CSE, Dr. C. V, RAMAN University, Bilaspur, India

M. Gupta et al. (eds.), *Computer Vision and Machine Intelligence in Medical Image Analysis*, Advances in Intelligent Systems and Computing 992,
https://doi.org/10.1007/978-981-13-8798-2_4

viewing time, and robustness of the methods for detection are few of the important challenges in the field of lung cancer diagnosis by Sinha and Patil [1]. Computer-aided diagnosis (CAD)-based methods are used in all modern research and diagnosis laboratories. False identification of cancerous elements is a common phenomenon in manual detection of the cancer by Patil and Kuchanur [2]. The automated approach is developed in CAD, wherein some image processing tools are also used in preprocessing, feature extraction, and postprocessing of the medical images. Few research papers on CAD are Chaudhary and Singh [3] and Hadavi et al. [4], and these papers emphasize on building an efficient and accurate automated approach for lung cancer identification.

We have implemented lung cancer detection using fuzzy inference system (FIS) to identify the prominent cancerous cells, including all four stages of CAD-like other research work. Fuzzy inference rules were applied to get accurate results for lung cancer detection.

Camarlinghi et al. [5] use CAD technique for identifying pulmonary nodules in CT scans. Abdullah and Shaharum [6] use feedforward neural network to classify lung nodules in X-ray images albeit with only a small number of features such as area, perimeter, and shape. The appropriate set of features and classification methods resulted in poor accuracy of CAD-based diagnosis results. Kuruvilla and Gunavathi [7] consider six distinct parameters including skewness, fifth and sixth central moments extracted from segmented single slices containing the highlighted features as suggested in Abdullah and Shaharum [6]. The feedforward backpropagation neural network is trained to evaluate accuracy for different features separately. This work claims a high detection rate of 88.5% with an average of 6.6 false positives (FPs) per CT scan on 15 CT scans but the size of database was not large. Hayashibe [8] implemented an automatic method on mass chest radiographs for the detection of new lung nodules. The marker locations are considered as regional minima and then the watershed algorithm is applied. Mori et al. [9] do the extraction of bronchus area from 3-D chest X-ray CT images. Initial finding of segmentation of CT scan using convolution neural network (CNN) classification was highlighted by Dwivedi et al. [10]. Alawaa et al. [11] suggest a new method for efficient training and detection of cancerous elements found in lung CT images. Current literature suggests that there is much work that has been already done in the field of lung cancer detection. However, the accuracy has not been attempted to improve with the help of fuzzy-based approaches.

2 Method, Results, and Discussion

Like other methods of CAD, the proposed method uses few important stages of implementation of lung cancer detection. We have used different sets of methods under each of the main stages, as described below.

Step 1: Read CT Image

Input CT image is read from databases of CT images of lung cancer images [National Cancer Institute, Lung Image Database Consortium (LIDC), and Image Database Resource Initiative (IDRI)] in JPEG formats. The image was described as having no nodule (no cancer cell), having nodule size $\leq$3 mm, and those having nodule size >3 mm. The image nodule size less than 3 mm has less probability of cancerous elements, and more probable cases are in size of more than 3 mm.

Step 2: Gabor Filter-based Enhancement

We used standard Gabor filter Sinha and Patil [1] for the enhancement of CT images that removes noise before subjecting them to subsequent stages of CAD.

Step 3: Thresholding-based Segmentation

Thresholding is most commonly used for segmentation task that first converts the grayscale image into binary image, and based on threshold value T, it assigns two levels to the images. A global value of threshold is selected and applied, which operates as estimating initial value of T, segmentation using T, separation of pixels greater and smaller than T, computation of average intensities (m_1 and m_2) and finally, we calculate new threshold value as

$$T_{new} = (m_1 + m_2)/2 \tag{1}$$

We go on checking id $|T - T_{new}| > \Delta T$; otherwise, recalculate average and determine new threshold by changing different values of T.

Step 4: Morphological Operation

Standard morphological operators of MATLAB are used, such as dilation and erosion. Binary lung mask is calculated with the help of four-connected neighborhood of all ones. The mask is multiplied with original lung CT images resulted from segmentation. Now, creating binary mask, we select region of interest (ROI) for further processing, as feature extraction.

Step 5: Feature Extraction

Texture features, local binary pattern, area, and perimeter are extracted and applied to fuzzy inference system for classification and detection of lung cancer. Haralick's texture features are used based on moments of joint probability density function (PDF), which is further estimated using co-occurrence of gray levels that are also known as gray level co-occurrence matrix (GCM) as discussed in Sinha and Patil [1]. These matrices are calculated with the help of unit pixel distance for different angles 0, 45, 90, and 135°. Finally, we get four sets of different GCM texture features. Local

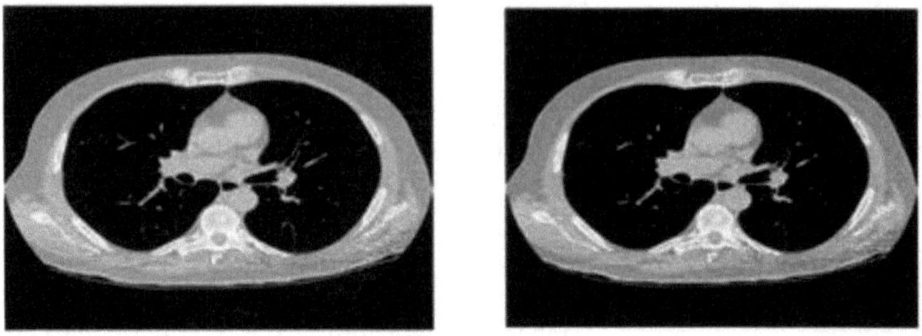

Fig. 1 Result of image enhancement using Gabor filter

Table 1 Comparison of enhancement methods

Samples (CT images)	PSNR (dB) in FFT	PSNR (dB) in Gabor filter
Sub1	2.0324	16.4880
Sub2	7.8653	13.4033
Sub3	3.3805	15.2433
Sub4	3.9786	14.4634
Sub5	7.5149	14.1895

binary pattern (LBP) by Sinha and Patil [1] is then computed. Area and perimeter of ROI are calculated as

$$\text{Area} = \sum (\text{White pixels in I(x, y)}) \tag{2}$$

$$\text{Perimeter} = \sum (\text{Boundary pixels in the image}) \tag{3}$$

Step 6: Fuzzy Inference System (FIS)

Fuzzy rules are applied for the identification of suspected regions in CT images. If fuzzy score is more than 70%, then the probability of suspected region is more; for less than 50%, the probability is less; and between 50 and 70%, the probability is moderate.

The result of image enhancement for one of the CT images, using Gabor filtering, is shown in Fig. 1 (original image is shown in left and the enhanced image is on right side) (Table 1).

Results of segmentation using watershed method, as thresholding method, can be seen in Fig. 2.

Now, CT images, both normal and abnormal, are processed using binarization and morphological operations as suggested in Sinha and Patil [1]. Figure 3 shows original image (left) and masked image (right) for normal image, and Fig. 4 shows original image (left) and masked image (right) for abnormal image.

Table 2 shows the values of white pixels and black pixels calculated upon the results of binarization operation.

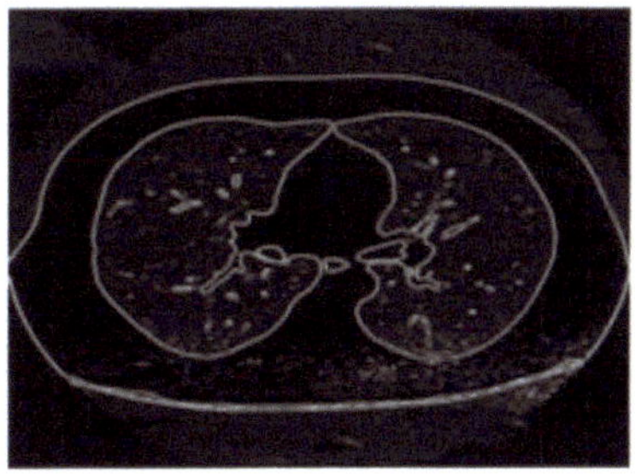

Fig. 2 Result of watershed segmentation method

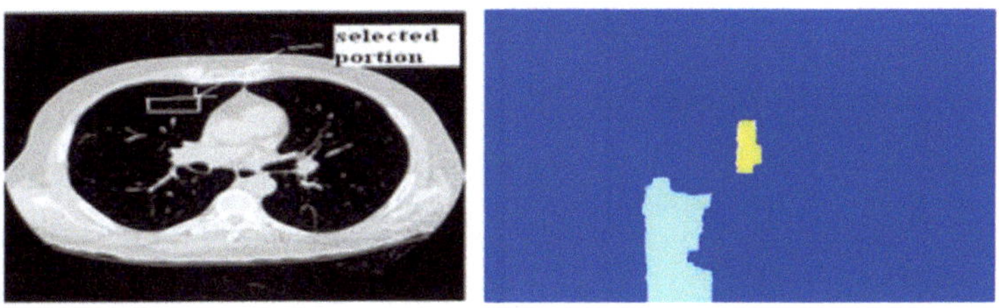

Fig. 3 Original images of normal lung and masked image

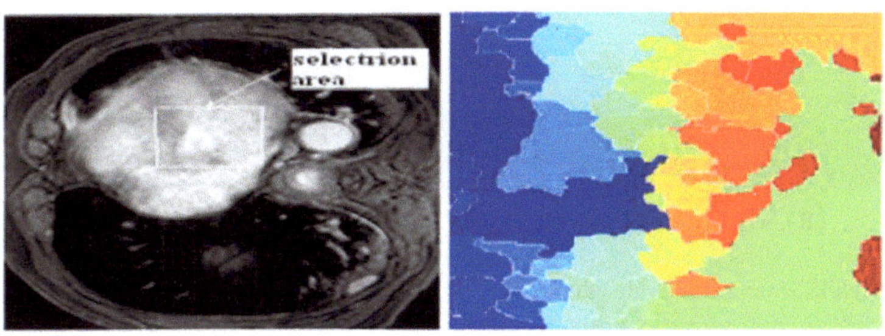

Fig. 4 Representation of abnormal lung image and masked image of lung

Table 2 Feature extraction after binarization operation

Normal image	No. of black pixels	No. of white pixels	Abnormal image	No. of black pixels	No. of white pixels
Subn1	169826	88238	Suban1	37431	65654
Subn2	174856	126275	Suban2	84733	111347
Subn3	158354	142777	Suban3	132155	181985
Subn4	179584	121547	Suban4	6449	22751
Subn5	67196	77565	Suban5	37441	64603

```
fis =
    name: 'anfis'
    type: 'sugeno'
    andMethod: 'prod'
    orMethod: 'max'
    defuzzMethod: 'wtaver'
    impMethod: 'prod'
    aggMethod: 'max'
    input: [1x2 struct]
    output: [1x1 struct]
    rule: [1x25 struct]
```

Fig. 5 Architecture of ANFIS algorithm

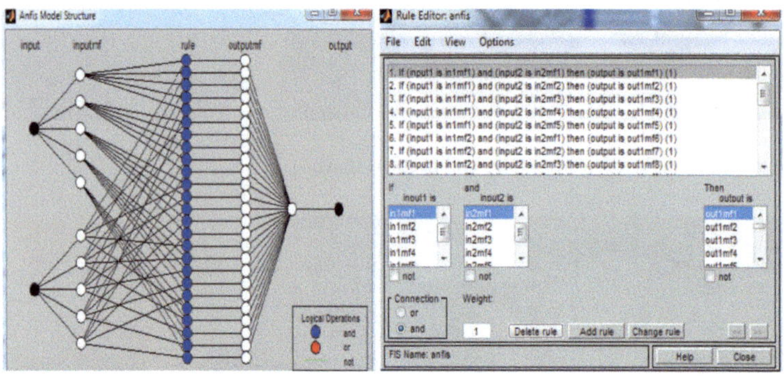

(a)ANFIS structure (b) Fuzzy rule editor

Fig. 6 ANFIS structure and rule editor

We counted number of black pixels for normal and abnormal cases and calculated average of the pixels. The average results in the threshold value for the classification of abnormal from normal. The number of white pixels is more in the abnormal lung images and accordingly, we can see in Table 2 that white pixels are more in case of abnormal images of various samples.

Finally, fuzzy inference system (FIS) employing the concept of neurons and implemented as artificial neural network FIS (ANFIS) and the architecture of ANFIS algorithm can be seen in Fig. 5.

Figure 6a represents the architecture of ANFIS and its rule editor executing various fuzzy rules that can be seen in Fig. 6b. ANFIS has 2 inputs, 25 fuzzy rules, and 1 output.

The ANFIS has 25 fuzzy rules, which are shown in Fig. 7.

Figure 8 shows the results of ANFIS, in which Fig. 8a depicts cancerous region and Fig. 8b shows noncancerous region in CT image of lung cancer. This classification

1. If (input1 is in1mf1) and (input2 is in2mf1) then (output is out1mf1) (1)
2. If (input1 is in1mf1) and (input2 is in2mf2) then (output is out1mf2) (1)
3. If (input1 is in1mf1) and (input2 is in2mf3) then (output is out1mf3) (1)
4. If (input1 is in1mf1) and (input2 is in2mf4) then (output is out1mf4) (1)
5. If (input1 is in1mf1) and (input2 is in2mf5) then (output is out1mf5) (1)
6. If (input1 is in1mf2) and (input2 is in2mf1) then (output is out1mf6) (1)
7. If (input1 is in1mf2) and (input2 is in2mf2) then (output is out1mf7) (1)
8. If (input1 is in1mf2) and (input2 is in2mf3) then (output is out1mf8) (1)
9. If (input1 is in1mf2) and (input2 is in2mf4) then (output is out1mf9) (1)
10. If (input1 is in1mf2) and (input2 is in2mf5) then (output is out1mf10) (1)
11. If (input1 is in1mf3) and (input2 is in2mf1) then (output is out1mf11) (1)
12. If (input1 is in1mf3) and (input2 is in2mf2) then (output is out1mf12) (1)
13. If (input1 is in1mf3) and (input2 is in2mf3) then (output is out1mf13) (1)
14. If (input1 is in1mf3) and (input2 is in2mf4) then (output is out1mf14) (1)
15. If (input1 is in1mf3) and (input2 is in2mf5) then (output is out1mf15) (1)
16. If (input1 is in1mf4) and (input2 is in2mf1) then (output is out1mf16) (1)
17. If (input1 is in1mf4) and (input2 is in2mf2) then (output is out1mf17) (1)
18. If (input1 is in1mf4) and (input2 is in2mf3) then (output is out1mf18) (1)
19. If (input1 is in1mf4) and (input2 is in2mf4) then (output is out1mf19) (1)
20. If (input1 is in1mf4) and (input2 is in2mf5) then (output is out1mf20) (1)
21. If (input1 is in1mf5) and (input2 is in2mf1) then (output is out1mf21) (1)
22. If (input1 is in1mf5) and (input2 is in2mf2) then (output is out1mf22) (1)
23. If (input1 is in1mf5) and (input2 is in2mf3) then (output is out1mf23) (1)
24. If (input1 is in1mf5) and (input2 is in2mf4) then (output is out1mf24) (1)
25. If (input1 is in1mf5) and (input2 is in2mf5) then (output is out1mf25) (1)

Fig. 7 Set of fuzzy rules

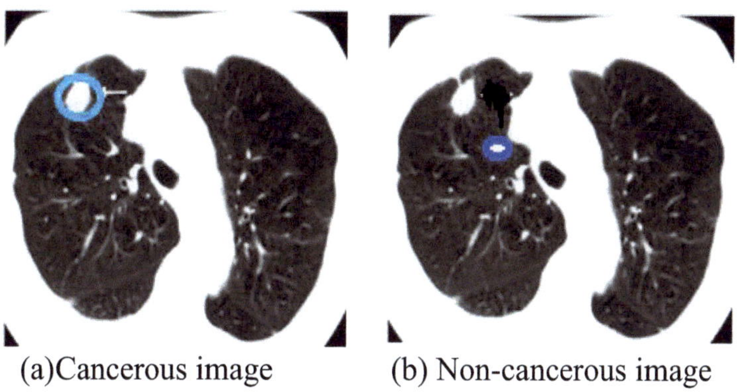

Fig. 8 **a** Cancerous image (area = 685.75 mm^2 and ANFIS score = 76.59%) and **b** Noncancerous image (area = 41.375 mm^2 and ANFIS score = 36.54%)

is based on area and score values, and ANFIS accordingly classifies the cancerous and noncancerous regions.

Methods/images	Existing methods [2, 4]		Proposed method	
	Area	Score	Area	Score
Sub1	305.25	40.45	305.25	43.45
Sub2	612	70.85	676	76.85
Sub3	500	70.79	524	75.79
Sub4	240.56	40.75	261.56	43.75
Sub5	345	40.8	363	43.8

3 Conclusions and Future Work

Lung cancer is said to have been one of the main causes of death. Automated system of diagnosis and analysis of lung cancer CT images, use CAD system. The CAD system used few important stages, namely, image acquisition, enhancement, segmentation, feature extraction, and classification. This paper suggests a framework of these stages, as a set of appropriate methods so that the lung cancer detection becomes efficient and diagnosis accuracy is improved. ANFIS has helped the classification significantly, and the separation of cancerous and noncancerous regions has been achieved. A framework of appropriate enhancement and segmentation methods, combined with ABFIS system, can greatly help the radiologists or physicians working toward lung cancer detection, especially at early stage. Robustness has always scope to be improved and therefore this would be further improved in future work.

References

1. Sinha, G.R., Patil S.: Medical Image Processing. Prentice Hall of India (2014)
2. Patil, D.S., Kuchanur, M.: Lung cancer classification using image processing. Int. J. Eng. Innov. Technol. (IJEIT) **2**(2), 55–62 (2012)
3. Chaudhary, A., Singh, S.S.: Lung cancer identification on CT images by using image processing. In: IEEE International Conference on Computing Sciences (ICCS). pp. 142–146 (2012)
4. Hadavi, N., Nordin, M., Shojaeipour, A.: Lung cancer diagnosis using CT-scan images based on cellular learning automata. In: The Proceedings of IEEE International Conference on Computer and Information Sciences (ICCOINS), pp. 1–5 (2014)
5. Camarlinghi, N., Gori, I., Retico, A., Bellotti, R., Bosco, P., Cerello, P., Gargano, G.E.L., Torres, R., Megna, M.Peccarisi, et al.: Combination of computer-aided detection algorithms for automatic lung nodule identification. Int. J. Comput. Assist. Radiol. Surg. **7**(3), 455–464 (2012)
6. Abdullah, A.A., Shaharum, S.M.: Lung cancer cell classification method using artificial neural network. Inf. Eng. Lett. **2**(1), 49–59
7. Kuruvilla, J., Gunavathi, K.: Lung cancer classification using neural networks for ct images. Comput. Methods Programs Biomed. **113**(1), 202–209 (2014)
8. Hayashibe, R.: Automatic lung cancer detection from X-ray images obtained through yearly serial mass survey. In: IEEE International Conference on Image Processing (1996)

9. Dwivedi, S.A., Borse, R.P., Yametkar, A.M.: Lung cancer detecton and classification by using machine learning and multinomial bayesian. IOSR J. Electron. Commun. **9**(1), 69–75 (2014)
10. WafaaAlawaa, Mahammad Nassef, Badr, Amr: Lung cancer detection and classification with 3D convolutional neural network (3D-CNN). Int. J. Adv. Comput. Appl. **8**(8), 409–417 (2017)
11. Armato, S.G., McLenman, G., Clarke, L.P.: National cancer institute, lung image database consortium (LIDC) and image database resource initiative (IDRI). **38**(2). 915–931
12. Cancer Statistics: Cancer J. Clin. 10–30 (2005)
13. Salman, N.: Image segmentation based on watershed and edge detection techniques. Int. Arab J. Inf. Technol. **3**(2), 104–110 (2006)
14. Mori, K., Kitasaka, T., Hagesawa, J.I, Toriwaki, J.I., et al.: A method for extraction of bronchus regions from 3D chest X-ray CT images by analyzing structural features of the bronchus. In: 13th International Conference on Pattern Recognition, pp. 69–77 (1996)
15. Raja, R., Sinha, T.S., Dubey, R.P.: Orientation calculation of human face using symbolic techniques and ANFIS. Publ. Int. J. Eng. Futur. Technol. **7**(7), 37–50. ISSN: 2455-6432 (2016)
16. Raja, R., Sinha, T.S., Dubey, R.P.: Recognition of human-face from side-view using progressive switching pattern and soft-computing technique. Assoc. Adv. Model. Simul. Tech. Enterp. Adv. B, **58**(1), 14–34. ISSN:-1240-4543 (2015)

Medical Image Compression Scheme Using Number Theoretic Transform

Salila Hegde and Rohini Nagapadma

Abstract In this paper, a new methodology is proposed for the medical image compression using number theoretic transforms or NTTs. NTTs are the discrete Fourier transforms carried over finite fields and rings. All the arithmetic operations are carried over a modulo number M. From the review of NTTs and their variants, it is found that NTTs involve only real integers, and the transform is reversible and hence no round-off errors in NTT-based algorithms. Another attractive feature is that NTTs of regular structures are also regular. These factors lay the foundation for the proposed lossless compression scheme of medical images. The variant of NTT known as Fermat number transform (FNT) is used for the proposed compression scheme as it involves less or no multiplications. The results obtained are favorable in terms of compression ratio and reduced number of computations. Further study and research is in progress to optimize the algorithm in terms of computations and hardware implementation of NTTs for real medical images. It is forecasted that with the use of dedicated hardware and optimization of these digital transforms, much higher compression ratio at faster speed may be achieved.

Keywords Number theoretic transform · Fermat number transform · Lossless compression · Mersenne number transform

1 Introduction

In the recent years, hospitals all over the world generate huge number of images of human body and internal organs for the reference of doctors and clinical specialists to diagnose and treat various diseases. A digital computer stores these images as

S. Hegde (✉)
NIE Institute of Technology, Mysore, India
e-mail: salilahegde@nieit.ac.in

R. Nagapadma
National Institute of Engineering, Mysore, India
e-mail: rohini_nagapadma@yahoo.co.in

M. Gupta et al. (eds.), *Computer Vision and Machine Intelligence in Medical Image Analysis*, Advances in Intelligent Systems and Computing 992,
https://doi.org/10.1007/978-981-13-8798-2_5

two-dimensional array of pixels of size M × N. An M × N binary image requires M × N × 1 bits, and grayscale image requires M × N × 8 bits. An RGB color image requires three times more storage than grayscale image of same size. Medical images like CT scan, MRI, etc., are volumetric in size, and hence, their storage requirements are significantly high. Bandwidth and speed are the issues to be considered as the transmission of medical images is vital due to telemedicine and increasing use of multimedia over the World Wide Web. These requirements make compression of medical images a desirable preprocessing operation before storage and transmission [1, 2].

Numerous image compression algorithms have been developed in the past decades and are in use. They can be generally classified as lossy and lossless algorithms. Lossy algorithms like JPEG and DWT-based algorithms give a large compression ratio but fail in preserving the image quality as the algorithms used are irreversible [3–8]. They also suffer from blocking artifacts and round-off errors. On the other hand, lossless algorithms perform better in preserving image quality as the algorithm is reversible, but at the compromise of lesser compression ratio. Lossless methods also require higher computational complexity [9, 10].

In this paper, a new lossless compression scheme is proposed by using a transform known as number theoretic transform. NTTs are the discrete Fourier transforms carried over finite fields and rings, and all the computations are modulo arithmetic operations [11–13]. NTTs are in use since 70s, and since then there have been many applications based on NTTs in the field of digital signal processing [14, 15]. There are variants of NTTs, namely, Fermat number transform, Mersenne number transform, pseudo-Fermat number transform, NMNT, etc., of which the former two transforms are found attractive as they involve lesser number of multiplications or no multiplications at all [16, 17]. As NTTs possess DFT like structures and properties, they are found useful for convolution, correlation, digital filtering, faster FFT calculation, etc. [18–21]. Even if there is a single entity change also in the given input, there is a drastic change in its transform if we use NTT. This property is found useful for efficient cryptographic applications [22, 23].

NTT of a regular structure is also regular and many coefficients are zero in the transform domain [24, 25]. NTTs involve only integers as all the arithmetic operations are carried over rings and fields over a modulo number M, and hence, NTT-based algorithms need not suffer from round-off errors as in JPEG compression. NTTs are highly reversible transforms. These features make NTT a very attractive option for lossless compression. The proposed medical image compression is based on this property. And also using FNT and by choosing a proper parameter for finding the transform, there can be less number of computations as in other transform cases. This paper is organized as follows. Section 2 reviews the definition of NTTs and its variants. Section 3 explains the image compression methodology using FNT. Sections 4 and 5 discuss implementation and results. Section 6 concludes with future scope.

2 Number Theoretic Transform

NTTs are discrete Fourier transforms carried over finite fields and rings. Transform methods are useful when the data can be processed in blocks. A general structure of the transforms is defined as

$$Y(K) = \sum_{n=0}^{N-1} y(n)\alpha^{nk} \tag{1}$$

It is found [26] that α is root of unity of order N, i.e., it has to satisfy the condition

$$\alpha^{N} = 1 \tag{2}$$

If this condition is satisfied, then the transforms will have property of CCP (circular convolution property). DFT is the transform which has CCP and α is replaced by exp(−j2π/N) [11]. In a finite field or ring of integers, all operations are carried out modulo M, where M is an integer. There exist a large class of transforms which have CCP. NTT belongs to such class. From the earlier work and literature on these transforms, here the definition for 1D transform is given as below:

$$X(k) = \sum_{n=0}^{N-1} x(n)\,\alpha^{n/k} \bmod M, k = 0,1, \ldots, N-1 \tag{3}$$

$$\begin{aligned} x(n) &= (N^{-1}\sum_{k=0}^{N-1} X(k)\alpha^{nk}) \bmod M, \\ n &= 0, 1, 2, \ldots, N-1 \end{aligned} \tag{4}$$

where M is some integer taken as a modulus. α is a root of unity of order N (i.e., $\alpha^{N} = 1 \bmod M$). N is transform length. x(n) and X(k) are the signal and its transform. It is desirable but not necessary that the modulus M is a prime number.

The 2D NTT and INTT are given by equations

$$X(k, 1) = \sum_{m=0}^{M-1}\sum_{n=0}^{N-1} X(m, n)\alpha_1^{mk}\alpha_2^{nl} \bmod F \tag{5}$$

$$\begin{aligned} x(m,n) &= (N^{-1}\sum_{k=0}^{N-1} X(k,l)\alpha_1^{mk}\alpha_2^{nk}) \bmod F \\ n &= 0, 1, 2, \ldots, N-1, m = 0, 1, 2, \ldots M-1 \end{aligned} \tag{6}$$

where F is the modulus; M, N are the dimensions of the 2D signal.

3 Image Compression Methodology Using NTTs

Fermat number transform (FNT) and Mersenne number transform (MNT) are the two significant variants of NTTs, where modulo M is given by F_t and M_p

$$F_l = 2^b + 1, \text{ where } \; b = 2^t, t = 0, 1, 2 \ldots . \tag{7}$$

$$M_p = 2^p - 1 \tag{8}$$

FNT is advantageous in terms of computational complexity as all multiplications can be replaced by shift operations by choosing α as 2 or multiples of 2. One more useful property of this transform is that FNT of a regular structure is also regular, and many coefficients will be zero. The foundation for image compression is that FNT can compress regular data effectively [24, 25]. Figure 1a is a test image of size 8 × 8 which has a periodic structure as Pr = 1, Pc = 4 where Pr and Pc are periodicities along row and columns, respectively. Figure 1b is the 2D FNT of Fig. 1a. As can be seen in the transform domain also, the coefficients appear in an order which has a relation with the period of original image, and many coefficients are zero.

Based on these properties of FNT, in this section, a medical image compression/decompression methodology is proposed. Figure 2a, b is the encoding and decoding models for the proposed medical image compression scheme.

1	2	3	4	1	2	3	4
1	2	3	4	1	2	3	4
1	2	3	4	1	2	3	4
1	2	3	4	1	2	3	4
1	2	3	4	1	2	3	4
1	2	3	4	1	2	3	4
1	2	3	4	1	2	3	4
1	2	3	4	1	2	3	4

(a)

20	13	0	30	0	25	0	19
0	0	0	0	0	0	0	0
0	0	0	0	0	0	0	0
0	0	0	0	0	0	0	0
0	0	0	0	0	0	0	0
0	0	0	0	0	0	0	0
0	0	0	0	0	0	0	0
0	0	0	0	0	0	0	0

(b)

Fig. 1 **a** Periodic image of size 8 × 8. **b** FNT of Fig. 1b

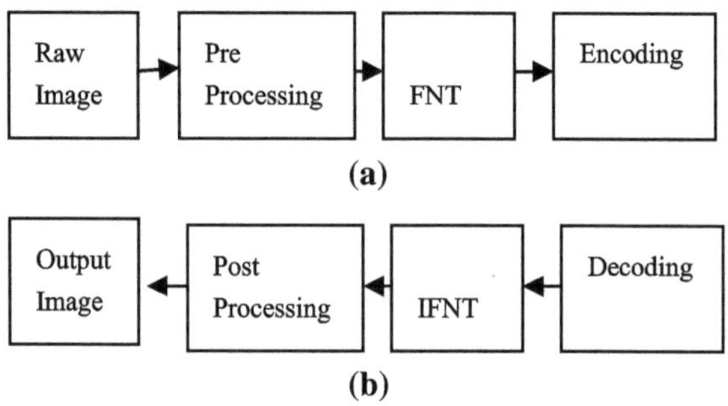

Fig. 2 **a** Encoding model. **b** Decoding model

3.1 Encoding Model

3.1.1 Raw Image

A raw medical image of size M × M is selected, and in the preprocessing stage, its 8 bit planes are separated and stored as eight binary images. In the next stage, FNT is carried out on these sets of binary images. And the result is encoded using RLE.

3.1.2 Preprocessing

In a gray scale image a pixel is represented using 8 bits. Upper bit planes carry most significant information compared to lower bit planes. It is observed that in a medical image as pixel redundancy is higher similarity between pixels in upper bit plane is higher. Pixels are more ordered in upper bit planes and have more regular image segments.

As FNT gives more number of zero coefficients on a regular image, the given gray scale image is segmented into its individual bit planes before applying FNT. Now the data element in these bit plane images has value either 0 or 1. Each bit plane image is of size M × M. Sub-images are formed using n number of consecutive bit planes to make them more regular. For example, the input raw image may be separated as four images of M × M size by grouping consecutive 2 bit planes. Ex: grouping bit plane 8 and 7. Similarly, two sub-images may be formed using upper 4 bit planes and lower 4 bit planes. This grouping is being done depending on the pixel correlation property.

3.1.3 FNT on Bit Plane Images

From the previous block, n numbers of sub-images using bit planes are obtained, which have more number of regular image segments. In this section, FNT is carried on all these n numbers of sub-images using 2D FNT. The given sub-image is divided into k numbers of blocks of 8 × 8 size. 2D FNT is applied in these blocks. If all coefficients of B (block) are either 0 or 1, 2D FNT of such blocks has $B_{0,0}$ element as 0 or a distinct value x. Hence, the block is encoded as 00 or 01 which saves 62 bits for a block. Otherwise, it is encoded as 10, and 1D FNT is carried on all eight rows of 8 × 8 blocks B and it is repeated. In 1D FNT also, the block is encoded as 00 or 01 or 10. If it is 10, then all 8 bits of that row are added to the encoded stream. This algorithm is carried out on all k blocks of sub-image, and the encoding stream is obtained. The following is the pseudocode for encoding and decoding:

3.1.4 Algorithm for FNT Encoding

Find total number of blocks k
For (j=1 to k)
{
Apply 2D FNT on Block Bj of size 8x8
If $Bj_{(0,0)}$ == 0 then encode it as 00
Else if $Bj_{(0,0)}$ == value in last column of Table 1 then encode it as 01
Else
{ Encode it as 10
for r= 1 to 8
{
Apply 1D FNT on rows of Block B i.e. B_{jr}
If $B_{jr(0,0)}$ == 0 then encode it as 00
Else if $B_{jr}(_{0,0})$ == value in last column of Table 1 then encode it as 01
Else Encode as 10 and append all 8 bits of $B_{j,r}$ to encoded stream } } }

Encoded bit stream E_1 is obtained, run-length encoding is performed, and final bit stream E2 is stored as compressed data.

3.2 Decoding Stage

Here the received data E_2 which is compressed is decoded to get back encoded bit stream E_1, which is further decoded to get back FNT coefficients. IFNT is applied on the decoded sub-images, and eight sub-images are obtained. In the post-processing stage, these sub-images are concatenated to get final gray scale decompressed image.

4 Implementation

In this section, the implementation details are discussed. All implementations are done using 64-bit processor on Windows-10 platform using MATLAB-13 version software. The following FNT parameters are chosen after studying the properties and verifying results using MATLAB code. Refer Table 1 for parameters for image block of different sizes.
The last column values are used to decode the FNT result as in algorithm explained in Sect. 3. Test images of different sizes used are shown in Figs. 4a, 5a, 6a, 7a and 8a. Results are obtained for these images by applying FNT on sub-images constructed using 1 bit plane, 2 bit planes, and 4 bit planes of input image. In case of 1 bit plane, FNT of 8 sub images are taken and the result is encoded; in case of 2 bit plane FNT is carried out on four sub-images; and in case of 4 bit planes, FNT is carried out on two sub-images using convenient FNT parameters. Compression ratio in each case is calculated as

Table 1 NTT parameters for different block sizes

Size of block (B)	Fermat number		Alpha	Result after FNT
N × N	t	Ft	α	Value of coefficients
16 × 16	0	3	4	All 1 s
16 × 16	1	5	6	All 1s
16 × 16	2	17	2, 6	B (0, 0) = 1, All other 0 s
16 × 16	3	257	2	B (0, 0) = 256, All other 0 s
8 × 8	0	3	4	All 1 s
8 × 8	1	5	6, 16	All 4 s
8 × 8	2	17	2, 8	B (0, 0) = 13, All other 0 s
8 × 8	3	257	4	B (0, 0) = 64, All other 0 s
4 × 4	0	3	4, 16	All 1 s
4 × 4	1	5	2, 8	B (0, 0) = 1, All other 0 s
4 × 4	1	5	6, 16	All 1 s
4 × 4	2	17	4	B (0, 0) = 16, All other 0 s
4 × 4	3	257	16	B (0, 0) = 16, All other 0 s

$$\text{Compression Ratio(CR)} = \text{Size in bits of original images/Size in bits of compressed image}$$

In the decompression process, IFNT is carried out on decoded sub-images, respectively. FNT forward and reverse transformation matrices are computed from Eqs. 5 and 6. For both forward and reverse transforms, direct transformation method is used. That is, $V = N.U.N^T$ where N is the transformation matrix, and U is the input image.

5 Results

Test images are taken from the World Wide Web. Images of different sizes are chosen based on their spatial frequency measure (SFM) [27] as in Table 2. SFM is defined as

$$\mathrm{SFM} = \sqrt{\mathrm{R}^2 + \mathrm{C}^2},$$

$$\mathrm{R} = \sqrt{\frac{1}{\mathrm{MN}} \sum_{j=1}^{M} \sum_{K=2}^{N} (f_{j,k} - f_{j,k-1})^2},$$

Table 2 SFM, CR, and compression time for different images

Image	Size	SFM	CR bit plane 8–1	Compression time(sec)
Angio.jpg	64 × 64	8.44	15–14.22	0.33
Skull.jpg	64 × 64	4.24	18–15	0.31
CTmono.jpg	128 × 128	9.008	18.61–15.7	0.043
CTort.jpg	128 × 128	6.86	24.97–14.22	0.037
CTL.jpg	128 × 128	3.82	30–14.42	0.032
Gd.jpg	256 × 256	5.07	49.95–14.22	0.097
Gd1.jpg	256 × 256	5.98	26.09–18.38	0.094
Image4.jpg	512 × 512	4.77	47.6–16.65	1.9
Image5.jpg	512 × 512	4.79	58–14.3	1.7

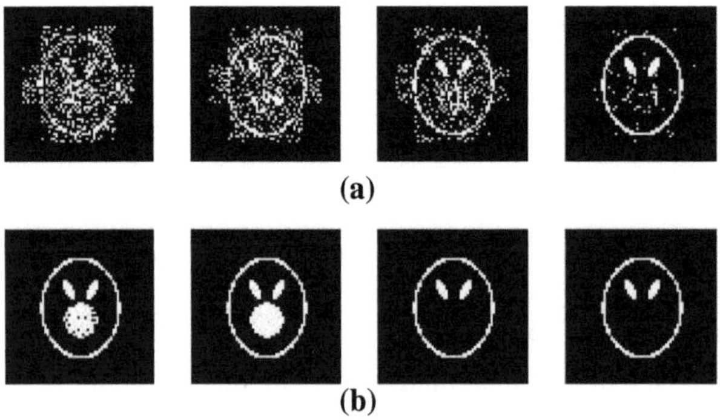

Fig. 3 **a** Bit plane 1 to bit plane 4 of skull.jpg. **b** Bit plane 5 to bit plane 8 of skull.jpg

$$C = \sqrt{\frac{1}{MN}\sum_{k=1}^{N}\sum_{j=2}^{M}(f_{j,k} - f_{j-1,k})^2} \tag{9}$$

where R is row frequency; C is column frequency; $f_{j,k}$ represents pixel at row and column position j, k; and M and N are row and column sizes of image.

An image with large SFM has less redundancy and less compression and vice versa. Each image is divided into its bit plane images. Figure 3a, b shows different bit planes of image skull.jpg.

FNT is carried out on each bit planes which is shown in Figs. 4b, 5b, 6b, 7b and 8b. Table 2 lists the SFM, compression ratio for different bit planes, and time taken in seconds. Compression ratio for eighth bit plane is maximum and first bit plane is minimum as higher bit planes have more redundancy than lower bit planes. Images are not shown according to actual size.

From this preliminary work carried out, it is estimated that the proposed methodology gives comparable compression ratio and time, and in some cases, it outperforms

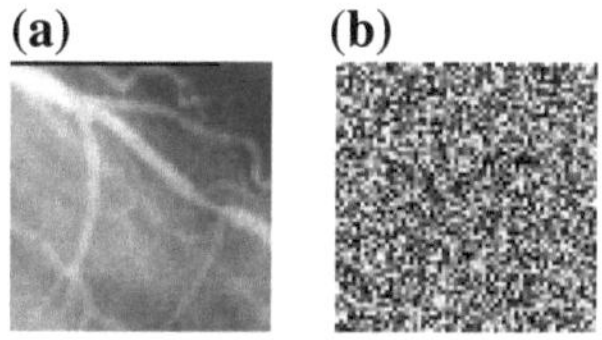

Fig. 4 **a** Angio.jpg size 64 × 64

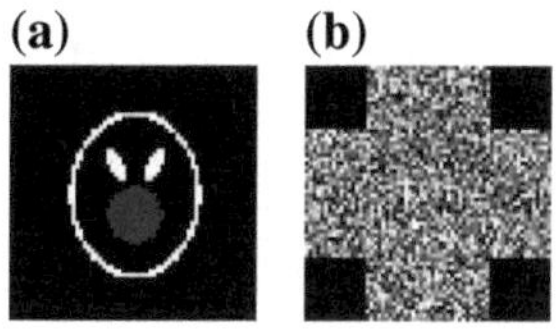

Fig. 5 **a** Skull.jpg size 64 × 64

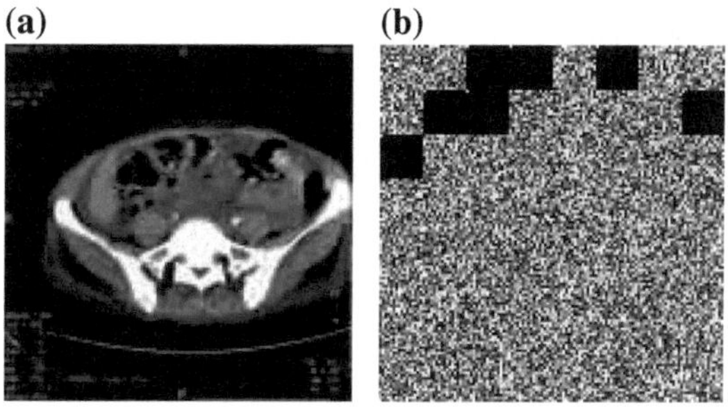

Fig. 6 **a** CTmono.jpg size 128 × 128

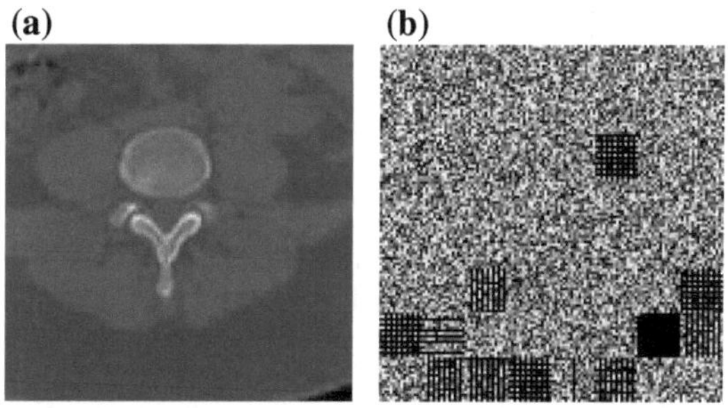

Fig. 7 **a** CtL.jpg size 128 × 128

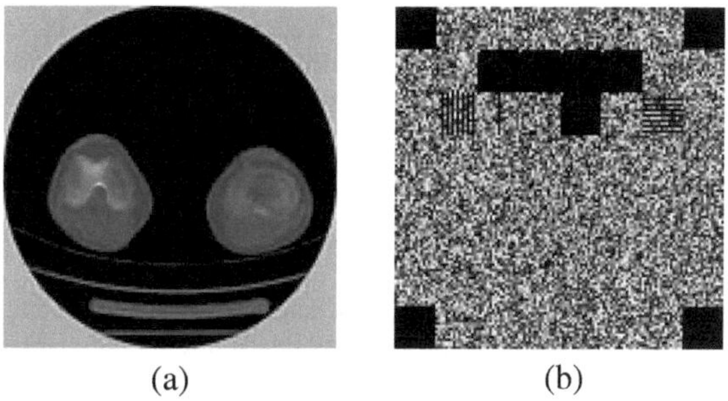

Fig. 8 **a** CTort.jpg size 128 × 128

other existing methods. Since number theoretic transforms are reversible transforms and totally error-free, the proposed methodology completely recovers the compressed image. And also by choosing $\alpha = 2$, all multiplications can be accomplished by simple shift operations and speed also enhanced.

6 Conclusion and Future Scope

In this research work, properties of number theoretic transforms and their variant FNT are verified, and a new compression scheme for medical images is implemented using Fermat number transform. The algorithm is applied to medical images of different sizes, and the obtained results are favorable in terms of compression time and ratio when compared to lossless methods JPEG-LS and JPEG 2000. Further work will be carried out to apply 1D FNT on the rows of a block if 2D FNT doesn't compress the N $\times$ N block. This is expected to give further improvement in compression ratio. With dedicated hardware using ASIC implementation and parallel computing paradigms, further improvements can be achieved [28, 29]. Hence, it is forecasted that it is worth developing NTT-based algorithms for medical images as they prove superior in terms of higher compression ratio and faster computation.

Acknowledgements This research is carried out at research Lab, NIE, Mysore. We would like to thank Dr. Narasimha Koulgoud, Professor, Department of ECE, NIE, Mysore, INDIA, for his valuable suggestions.

References

1. Jain, A.K.: Fundamentals of Digital Image Processing, 2nd edn
2. Jain, A.K.: Image data compression: a review. Proc. IEEE **69**, 349–389 (1981)
3. Useitch, B.: A tutorial on modern lossy wavelet image compression, foundations of JPEG 2000, IEEE Signal Process. Mag. (2001)
4. Wallace, G.: The JPEG still picture compression standard. IEEE Trans. Consum. Electron. (1991)
5. Xiong, Z., Ramchandran, K., Orchard, M.T., Zhang, Y.-Q.: A comparative study of DCT and wavelet based image coding. IEEE Trans. Circuits Syst. Video Technol. **9**(5) (1999)
6. Said, A., Pearlman, W.: A new fast and efficient image codec based on set partitioning in hierarchical trees. IEEE Trans. Circuits Syst. Video Technol. (1996)
7. Wohlberg, B., de Jager, G.: A review of the fractal image coding literature. IEEE Trans. Image Process. **8**(12) (1999)
8. Iano, Y., da Silva, F.S., Cruz, A.L.M.: A fast and efficient hybrid fractal-wavelet image coder. IEEE Trans. Image Process **15**(1) (2006)
9. Miaou, S.-G., Keand, F.-S., Chen, S.-C.:A lossless compression method for medical image sequences using JPEG—LS and interframe coding. IEEE Trans. Inf. Technol. Bio Med. **13**(5) (2009)
10. Rabbani, M., Joshi, R.: An overview of the JPEG2000 still image compression standard. Signal Process. Image Commun. (2002)
11. Pollard, J.M.: The fast fourier transform in a finite field. Math. Comput. **25**, 365–374 (1971)

12. Rabiner, L.R., Gold B.: Theory and Application of Digital Signal Processing, pp. 419–433. Prentice-Hall, London (19751)
13. Agarwal, R.C, Cooley, J.W.: New algorithms for digital convolution. IEEE Trans. ASSP-25 106–124 (1977)
14. Gudvangen, S., Buskerud, H.: Practical Applications of Number Theoretic Transforms. AUI, EMR, Kongsberg, Norway
15. Siu, W.C.: Number theoretic transform and its applications to digital signal processing. In: Proceedings of IERE HK Section Workshop on Adv. Micro, and DSP, Hong Kong, pp. 76–10 (1982)
16. Agarwall, R.C., Burrus, C.S.: Fast convolution using Fermat number transform with application to digital filtering. ibid. ASSP-22, pp. 87–97 (1974)
17. Rader, C.M., Discrete convolution via Mersenne transform. IEEETrans. **C-21**, 1269–1273 (1972)
18. Agarwal, R.C., Burrus, C.S.: Number theoretic transforms to implement fast digital convolution. In: Proceedings of the IEEE, vol. 63, pp. 550–560 (1975)
19. Veg, E., Leibowitz, L.M.: Fast complex convolution using number theoretic transforms, p. 7935. Naval Research Lab, Washington, DC (1975)
20. Toivonen, T., Heikkila, J.: Video filtering with Fermat number theoretic transforms using residue number system. IEEE Trans. Circuits Syst. Video Technol. **16**(1) (2006)
21. Conway, R.: Modified overlap technique using Fermat and Mersenne transforms. IEEE Trans. Circuits Syst.-II: Express Briefs, **53**(8) (2006)
22. Tamori, H.: Asymmetric fragile watermarking using a number theoretic transform. IEICE Trans. Fundam. **E92-A**(3) (2009)
23. Pedrouzo-Ulloa, A., Troncoso-Pastoriza, J.R., Pérez-González, F.: Number theoretic transforms for secure signal processing. IEEE Trans. Info. Forensics Secur. **12**(5) (2017)
24. Boussakta, S. Shakaff, A.Y., Marir, F., Holt, A.G.J.: Number theoretic transforms of periodic structures and their applications. In: IEE Proceedings, vol. 135. Pt. G (1988)
25. Hegde, S., Nagapadma, R.: Lossless compression scheme for regular images. IJCA **158**(9), 7–12 (2017)
26. Nicholson, P.J.: Algebraic theory of finite fourier transforms. J. Comput. Syst. Sci. **5**, 524–527 (1971)
27. Eskicioglu, M., Fisher, P.S.: Image quality measures and their performance. IEEE Trans. Commun. **43**(12), 2959–2965 (1995)
28. Truong, T.K., Yeh, C.S., Reed, I.S., Chang, J.J.: VLSI design of number theoretic transforms for a fast convolution. In: Proceedings of the IEEE International Conference on Compute Design, VLSl in Computing, ICCD, New York. U.S.A, pp. 200–203 (1983)
29. Towers, P.J., Pajayakrit, A., Holt, A.G.J.: Cascadable NMOS VLSl circuit for implementing a fast convolution using Fermat number transforms. In: IEE Proceedings of the G Electronic Circuits and Systems, vol. 134, no. 2, pp. 57–66 (1987)

The Retinal Blood Vessel Segmentation Using Expected Maximization Algorithm

R. Murugan

Abstract The Retinal Blood Vessel segmentation plays a vital role in automatic retinal disease screening systems. It helps in the screen process of glaucoma, diabetic retinopathy, and other eye- related diseases. The primary objective of this paper is to consequently segment the vessels in fundus retinal pictures, which encourages us in diabetic retinopathy screening. The initial enhancement of image is carried out using Histogram Equalization. After which, the green channel of the image is applied with morphological image processing to remove the optic disc. The image segmentation is then performed to modify the intensity of contrast and little pixels viewed as noise are evacuated. The obtained image would represent the blood vessels of the original image. This paper proposes an expected maximization algorithm to segment the blood vessels in the human retina. The novelty of these method is to perform uniform intensity distribution in retinal images. The proposed method was tested by publicly available datasets such as DRIVE, STARE, MESSIDOR, DIARETDB0, and University of Lincoln. The proposed method has obtained an average area under receiver operating characteristics of 0.9203. Moreover, this shows a better performance than other state-of-the-art methods.

Keywords Retina · Blood vessels · Diabetic retinopathy · Histogram equalization · Optic disc · Morphological processing

1 Introduction

The Blood Vessel Segmentation (BVS) is a fundamental advance in the medicinal finding of retinal fundus pictures as it helps in the determination of visual infections like Diabetic Retinopathy (DR). DR is an intricacy of diabetes and is a noteworthy reason for visual deficiency in created nations. The patients probably won't see lost

R. Murugan (✉)
Department of Electronics and Communication Engineering,
National Institute of Technology Silchar, Assam 788010, India
e-mail: murugan.rmn@gmail.com

M. Gupta et al. (eds.), *Computer Vision and Machine Intelligence in Medical Image Analysis*, Advances in Intelligent Systems and Computing 992,
https://doi.org/10.1007/978-981-13-8798-2_6

vision, until the point that it turned out to be excessively extreme, henceforth early determination and opportune treatment is fundamental to defer or anticipate visual impede and even visual impairment. The BVS can improve screening for retinopathy by lessening the quantity of false positive [1]. It chiefly influences about 80% of patients experiencing diabetics for a long time or more [2]. DR is a dynamic infection, which can progress from mellow stage to the proliferative stage. For the most part, DR is characterized into two principal stages, to be specific Non-Proliferative Diabetes Retinopathy (NPDR) and Proliferative Diabetic Retinopathy (PDR). NPDR is further classified to mild, moderate, or severe stages. DR mainly occurs due to changes in blood vessels of retina, as the blood glucose level increases. The walls of the blood vessels will then become weak and start to break and leaks the blood around them. Afterward, the cells in the retina will bite the dust from an absence of sustenance and the vessels may quit conveying blood for all time. Accordingly, old vessels do not work appropriately and new but rather unusual ones will develop to take their place. They cannot support the retina legitimately and may develop into the straightforward internal piece of the eye, and further influence vision [3]. However, manual BVS isn't reasonable on the grounds that the vessels in a retinal picture are mind boggling and have low contrast [4]. This paper displays a computerized BVS in the human retina. The remaining part of the paper is organized as follows. The related works are presented in Sect. 2. Section 3 presents the methodology. The results and conclusions are discussed in Sects. 4 and 5, respectively.

2 Related Works

This literature analysis provides the most generally proposed vessel segmentation algorithms to segment blood vessels in the retinal fundus images [2]. In vessel segmentation algorithms the input image first undergoes a preprocessing step, which typically concerns noise suppression, data normalization, contrast enhancement and conversion of color image to grayscale image. The necessary of preprocessing steps is to characterize the image features from various image modalities such as contrast, resolution, noise, etc. This literature survey presents an overall workflow of existing vessel segmentation algorithms. Furthermore, the existing segmentations are categorized into four different sections such as vessel enhancement [2], machine learning [2], deformable models [2], and tracking of vessels [2]. Through vessel enhancement approaches, the quality of vessel perception is improved, e.g. by increasing the vessel contrast with respect to background and other noninformative structures. A strong and established literature on vessel enhancement approaches already exists. Examples include matched filtering [5], vessels-based approaches [6], Wavelet [7], and diffusion filtering [8]. These methods provide the optimal resolution and contrast in both time and frequency domains. There are two principal classes of machine learning approaches: unsupervised [9] and supervised [10]. The former finds a model able to describe hidden arrangement of input image-derived features, with- out any prior knowledge or supervision, while the latter learns a data model from a set of already labelled features. The deformable models think about bends or surfaces, characterized inside the picture space, that can move and disfigure affected by interior and outside forces. The previous is intended to keep smooth amid the disfigurement while the last is drawn in toward the vessel limit. Deformable model methodologies can be isolated in edge-based [11] and region- based [12]. Blood vessel tracking algorithms usually consist in the definition of seed points followed by a growth process guided by image-derived constraints. Seed focuses can be either physically characterized or gotten through vessel upgrade approaches. The following methodologies are especially valuable to fragment-associated vascular trees, for which the division can be accomplished utilizing a predetermined number of seed focuses. The tracking methods can be isolated in template matching [13], model based [14], finding minimum cost path [15], and cost function estimation [16]. In view of the survey, distinctive BVS methods are right now utilized in DR screening and a proper decision of the segmentation algorithm is compulsory to manage BVS attributes such as resolution, noise, and vessel contrast. Hence, this paper has introduced to adopted the BVS attributes.

3 Methodology

The BVS is one of the key processing in DR screening. The separation of the fundus image will when all is said in done, be splendid at within and decline along the edge, consequently pre-processing is essential to restrain this effect and have a continuously uniform image. After which, the green channel of the image is applied with morphological image processing to remove the optic disc. The segmentation is then performed to change the difference force and little pixels viewed as clamor are expelled. Another green channel picture is prepared with segmentation and joined with the veil layer. These two pictures are analyzed and the distinctions are expelled. The obtained image would represent the blood vessels of the original image. The proposed methodology is shown in Fig. 1. This section discusses in greater detail of the extraction of the blood vessels. The fundus image is first preprocessed to standardize its size to 576 × 720. The intensity of the green channel is then inverse before adaptive histogram equalization is applied.

3.1 Green Component Estimation

The vessel, by and large, has cut down reflectance differentiated and establishment of the retina, the green component estimation was used in the examination and it shows the best multifaceted nature between the vessels and the establishment of the retina. The vessels seem more differentiated in the green band than in the red or blue groups, subsequently, the green part of the picture is utilized. The green component of retinal image will have the black and white fundus image and it retains its color

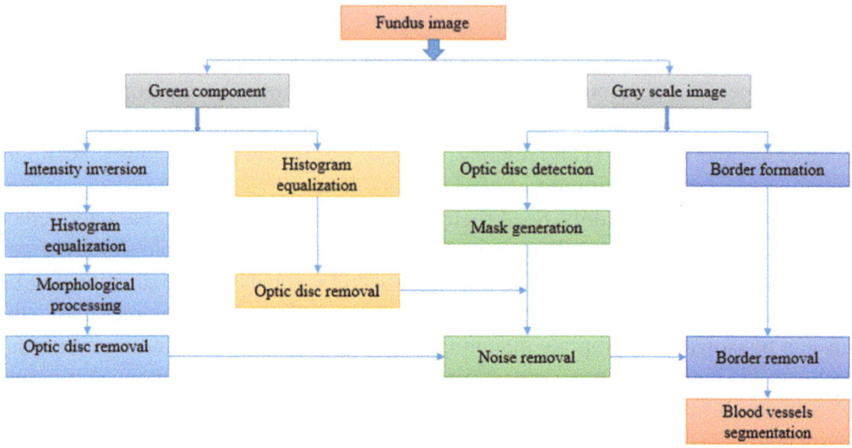

Fig. 1 Proposed methodology

property throughout the image process. In the next step, the green component image is inverse. Here, the white image is complemented to black image and black image to white image. Thus, it gives more clear identification of Blood Vessels.

3.2 Histogram Equalization

The fundus image often contains background intensity variation due to nonuniform illumination. In background pixel of the retinal image, there will be different intensities in the same image. In order to make the image contrast equal, Contrast Adaptive Histogram Equalization is used (CAHE). The CAHE is a technique used for the contrast improvement in the image. This method differs from ordinary histogram methods, where this method computes several histograms and each identifies lightness values of retinal image.

3.3 Morphological Operations

Morphological tasks are a lot of picture preparing activities that dissects the shapes inside the picture. It applies an organizing component to the information picture and yield picture of a similar size. The yield estimation of every pixel is controlled by the neighboring pixels with its relating pixel of image information. The size and state of the organizing component influence the number of pixels being included or expelled from the article in the image. The fundamental morphological tasks, for example, dilation, erosion, and opening are utilized in this progression.

3.4 Optic Disc Removal

The Optic Disc is a circular patch in the posterior part of eye. It can be removed from the retinal fundus image by subtracting the morphological opening applied image from adaptive histogram equalization applied image.

3.5 Border Detection

The border formation is to clean off the noisy edges. Here, two strategies are utilized in the discovery of roundabout outskirt of the image. The grayscaling picture rather than the green station is utilized as it is progressively productive in fringe identification. The principal strategy utilizes a vigilant technique to identify the edges previously encasing the round locale with a best and base bar. The border detection is gotten in the wake of subtracting the expanded image with the disintegrated image [2]. It is enacted when a boisterous picture is acquired rather than a roundabout fringe. This technique inverses the force of the picture first before picture division is connected. The roundabout area is filled accordingly and the round fringe is gotten subsequent to subtracting the expanded picture with the dissolved image. The last veins are divided after the expulsion of the roundabout outskirt.

3.6 Blood Vessel Segmentation

The blood vessel is segmented by the proposed Expected Maximization (EM) algorithm. The EM algorithm was derived to correct intensity inhomogeneity for bias fields correction of Magnetic Resonance Images (MRI) brain images by Neil et al. [17]. Then, the EM algorithm was developed to segment MR image by Leemput et al. [18]. The limitation of EM algorithm is that intensity distribution in brain image. This methodology presents EM algorithm to segment blood vessels in the human retina. The intensity distribution is addressed by the preprocessing step. The EM segmentation algorithm is more sensitive than K-means and Fuzzy c- means algorithms.

The EM is an optimization method (Shown in Eq. 1) to estimate unknown parameters θ, given measurement data U and unknown hidden nuisance variable J. In specific cases the posterier probability parameters θ may be maximize by the given data U, marginalizing hidden nuisance variable J. Here, the unknown parameter is blood vessels, the measured data is border of the image, and the hidden variable is optic disc.

$$\theta = argmax_{\theta} \sum_{J \epsilon \xi} \left(\theta, \frac{J}{U} \right) \tag{1}$$

$x_1, x_2, x_3 \ldots x_n$ and model p(x,z) where z is the latent variable.

$$l(\theta) = \sum_{i=1}^{m} \log p(x_i, \theta) \tag{2}$$

$$l(\theta) = \sum_{i=1}^{m} \log \sum_{z} p(x_i, \theta) \tag{3}$$

The log-likelihood is described in terms of x, z, and θ. But z the latent variable is unknown, and for finding these unknown variables, the approximations has taken place. E step for each i

$$\theta_i(z^i) = p\left(\frac{z^i}{x^i}, \theta\right) \tag{4}$$

M step for all z

$$\theta = argmax_{\theta} \sum_{i} \sum_{z^i} \theta_i z^i \log \frac{p(x_i, z^{i,}\theta)}{\theta_i z^i} \tag{5}$$

where θ_i is the posterior distribution of z^i given the x^i Finally, the unknown variable θ that is blood vessels are segmented.

4 Results

4.1 Dataset

The database used for blood vessel segmentation is shown in Table 1. Thus, a set of 1550 retinal pictures is tried in which 679 pictures are ordinary and 871 pictures are strange for a mechanized segmentation of vessels in the retina.

A MATLAB (2017b) prototype was used to run the algorithm developed for the extraction of blood vessels in retinal images for each image on a laptop (2.50 GHz Intel(R) Core(TM) i5-2450M CPU and 4.00 GB RAM). First, as the contrast of the fundus image tends to be bright at the center and diminish at the side, hence preprocessing is applied to minimize this effect and have a more uniform image. After which, the green channel of the image is applied with morphological image processing to remove the optic disc. The segmentation is then performed to modify the difference in force and little pixels viewed as commotion are removed. Another green channel picture is handled with picture division and joined with the cover layer. These two pictures are analyzed and the distinctions are expelled. The acquired picture would speak to the veins of the first picture. Figure 2 demonstrates the well-

Table 1 Database used for vessel segmentation

S.No	Dataset	Total images	Normal images	Abnormal images
1	STARE	81	31	50
2	DRIVE	40	33	7
3	MESSIDOR	1200	546	654
4	DIARETDB0	130	20	110
5	UNIVERSITY OF LINCOLN	99	49	50
	Total	1550	679	871

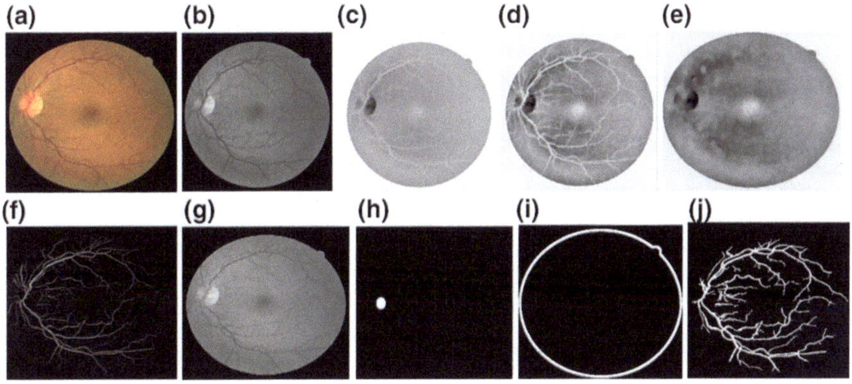

Fig. 2 **a** Input image. **b** Green component image. **c** Intensity inversion image. **d** Histogram equalized image. **e** Morphological image. **f** Optic disc removed image. **g** Grayscale image. **h** Optic disc detected image. **i** Border removed image. **j** Blood vessel segmented image

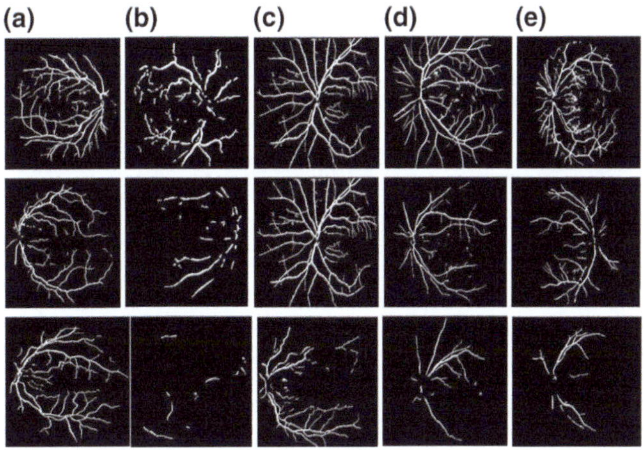

Fig. 3 **a** DRIVE. **b** STARE. **c** MESSIDOR. **d** DIARETDB0. **e** University of Lincoln

ordered yield of the proposed EM calculation and Fig. 3 demonstrates the example of pictures of utilized datasets.

The performance of the proposed blood vessel segmentation method was evaluated by means of the Receiver Operating Characteristic (ROC) curve stated by University of Nebraska, Medical Center, Department of Internal Medicine. The ordinate of the ROC curve is the sensitivity, which is the capacity of the technique to characterize the irregular pictures and the abscissa is identified with the explicitness, which is the capacity to arrange the typical pictures. The values of the sensitivity and the specificity have been obtained by Eqs. 6 and 7, respectively, the value of the area under the ROC curve (AUC)is described using dice similarity coefficient by Eq. 8.

$$Sensitivity = \frac{TruePositive}{TruePositive + FalseNegative} \tag{6}$$

$$Specificity = \frac{TrueNegative}{TrueNegative + FalsePositive} \tag{7}$$

$$AUC = \frac{2TruePositive}{FalsePositive + FalseNegative + 2TruePositive} \tag{8}$$

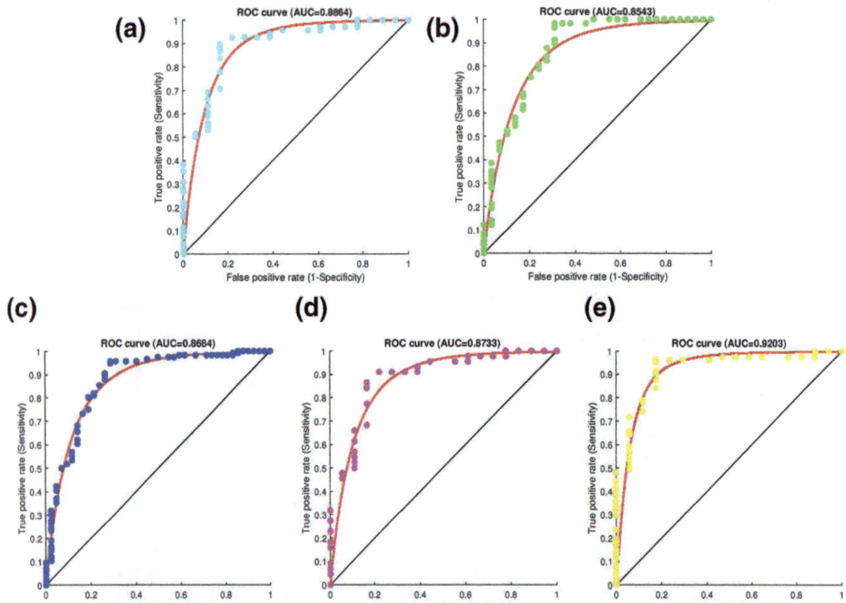

Fig. 4 Roc curve for various datasets. **a** STARE. **b** DRIVE. **c** MESSIDOR. **d** DIARETDB0. **e** University of Lincoln

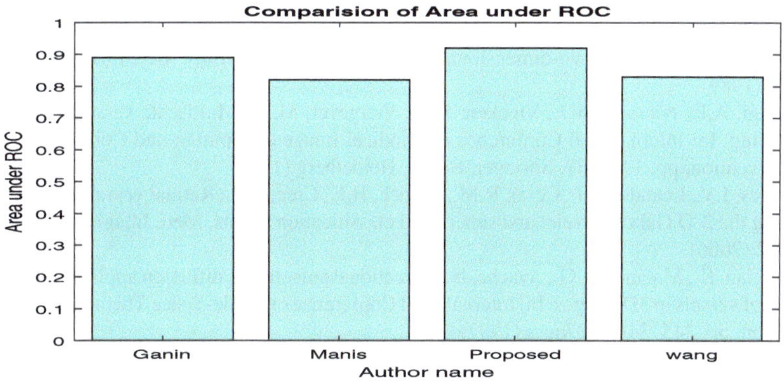

Fig. 5 Comparison of area under ROC in DRIVE dataset

The AUC curve was generated for tested datasets that can be found in Fig. 4. The obtained AUC value is compared with other state-of-the-art methods, proposed by Ganien et al. [19], Manis et al. [20] and Wang et al. [21], out of which the EM algorithm has produced better values that can be found in Fig. 5.

5 Conclusion

This paper proposes a novel blood vessel segmentation algorithm for retinal fundus images. The proposed framework deployed several fundamental image processing techniques. The backbone of this algorithm is newly proposed for blood vessel segmentation. The experimental results demonstrate that proposed algorithm outperforms the state-of-the-art approaches in terms of area under ROC curves over the five publicly available datasets such as STARE, DRIVE, MESSIDOR, DIARETDB0, and University of Lincoln.

References

1. Jiang, Z., Zhang, H., Wang, Y., Ko, S.B.: Retinal blood vessel segmentation using fully convolutional network with transfer learning. Comput. Med. Imaging Graph. **68**(1), 1–5 (2018)
2. Moccia, S., De Momi, E., El Hadji, S., Mattos, L.S.: Blood vessel segmentation algorithms review of methods, datasets and evaluation metrics. Comput. Methods Programs Biomed. **158**(1), 71–91 (2018)
3. Yue, K., Zou, B., Chen, Z., Liu, Q.: Improved multi-scale line detection method for retinal blood vessel segmentation. IET Image Process. **12**(8), 1450–1457 (2018)
4. Zhao, J., Yang, J., Ai, D., Song, H., Jiang, Y., Huang, Y., Zhang, L., Wang, Y.: Automatic retinal vessel segmentation using multi-scale superpixel chain tracking. Digit. Signal Process. **81**(1), 26–42 (2018)

5. Chaudhuri, S., Chatterjee, S., Katz, N., Nelson, M., Goldbaum, M.: Detection of blood vessels in retinal images using two-dimensional matched filters. IEEE Trans. Med. imaging **8**(3), 263–269 (1989)
6. Frangi, A.F., Niessen, W.J., Vincken, K.L., Viergever, M.A.: Multiscale vessel enhancement filtering. In: International Conference on Medical Image Computing and Computer-Assisted Intervention, pp. 130–137. Springer, Berlin, Heidelberg (1998)
7. Soares, J.V., Leandro, J.J., Cesar, R.M., Jelinek, H.F., Cree, M.J.: Retinal vessel seg- mentation using the 2-D Gabor wavelet and supervised classification. Trans. Med. Imaging **25**(9), 1214–1222 (2006)
8. Krissian, K., Malandain, G., Ayache, N.: Directional anisotropic diffusion applied to segmentation of vessels in 3D images. In: International Conference on Scale-Space Theories in Computer Vision, pp. 345–348. Springer (1997)
9. Hassouna, M.S., Farag, A.A., Hushek, S., Moriarty, T.: Cerebrovascular segmentation from TOF using stochastic models. Med. Image Anal. **10**(1), 2–18 (2006)
10. Nekovei, R., Sun, Y.: Back-propagation network and its configuration for blood vessel detection in angiograms. IEEE Trans. Neural Netw. **6**(1), 64–72 (1995)
11. Xu, C., Pham, D.L., Prince, J.L.: Image segmentation using deformable models. In: Handbook of Medical Imaging. pp. 129–174 (2000)
12. Chan, T., Vese, L.: An active contour model without edges. In: International Conference on Scale-Space Theories in Computer Vision, pp. 141–151. Springer, Berlin, Heidelberg (1999)
13. Carrillo, J.F., Hoyos, M.H., Dvila, E.E., Orkisz, M.: Recursive tracking of vascular tree axes in 3D medical images. Int. J. Comput. Assist. Radiol. Surg. **1**(6), 331–339 (2007)
14. Friman, O., Hindennach, M., Khnel, C., Peitgen, H.O.: Multiple hypothesis template tracking of small 3D vessel structures. Med. Image Anal. **14**(2), 160–171 (2010)
15. Cohen, L.D., Kimmel, R.: Global minimum for active contour models: A minimal path approach. Int. J. Comput. Vis. **24**(1), 57–78 (1997)
16. Deschamps, T., Cohen, L.D.: Fast extraction of minimal paths in 3D images and applications to virtual endoscopy. Med. Image Anal. **5**(4), 281–299 (2001)
17. Trevor, H., Tibshirani, R., Friedman, J.: Unsupervised learning. In: The Elements of Statistical Learning, pp. 485–585. Springer, New York, NY (2009)
18. Leemput, V., Maes, F.: Automated model-based tissue classification of MR images of the brain. IEEE Trans. Med. Imaging **18**(10), 897–908 (1999)
19. Ganin, Y., Lempitsky, V.S.: N4-fields: neural network nearest neighbor fields for image transforms, In: Asian Conference on Computer Vision, pp. 536–551. Springer, Cham (2014)
20. Maninis, K.K., Pont-Tuset, J., Arbelez, P., Van Gool, L.: Deep retinal image understanding. In: Ourselin, S., Joskowicz, L., Sabuncu, M., Unal, G., Wells, W. (eds) Medical Image Computing and Computer-Assisted Intervention, Lecture Notes in Computer Science, vol. 9901, pp. 140–148. Springer, Cham (2016)
21. Wang, S., Yin, Y., Cao, G., Wei, B., Zheng, Y., Yang, G.: Hierarchical retinal blood vessel segmentation based on feature and ensemble learning. Neurocomputing **149**(1), 708–717 (2015)

Classification Algorithms to Predict Heart Diseases—A Survey

Prakash Ramani, Nitesh Pradhan and Akhilesh Kumar Sharma

Abstract It has been observed that the deaths due to heart disease are increasing day by day, and therefore, the need arises for predicting cardiovascular diseases beforehand. By predicting heart disease in advance, one can start the treatment at an early stage and avoid life-threatening situations. In this research article, models using various classification algorithms are generated, and prediction accuracy is compared. Artificial neural network has the highest prediction accuracy among the other algorithms, viz., K-Nearest Neighbors (KNN), Decision Tree (DT), Support Vector Machine (SVM), and Gaussian Naïve Bayes (GNB), and Artificial Neural Network (ANN) is used for generating models.

Keywords K-nearest neighbors · Decision tree · Support vector machine and Gaussian Naïve Bayes · Artificial neural network · Prediction · Accuracy

1 Introduction

Heart disease refers to problems associated with the heart. According to the Centers for Disease Control (CDC), the major cause of death in the United Kingdom, United States, Canada, and Australia is heart disease. In United States, one in every four deaths is due to cardiovascular diseases [1].

Congenital heart disease, arrhythmia, coronary artery disease, dilated cardiomyopathy, myocardial infarction, heart failure, hypertrophic cardiomyopathy, mitral regurgitation, mitral valve prolapse, and pulmonary stenosis are different types of

---

P. Ramani · N. Pradhan (✉) · A. K. Sharma
Manipal University Jaipur, Dehmi Kalan, Jaipur, India
e-mail: nitesh.pradhan@jaipur.manipal.edu

P. Ramani
e-mail: Prakash.ramani@jaipur.manipal.edu

A. K. Sharma
e-mail: Akhileshkumar.sharma@jaipur.manipal.edu

M. Gupta et al. (eds.), *Computer Vision and Machine Intelligence in Medical Image Analysis*, Advances in Intelligent Systems and Computing 992,
https://doi.org/10.1007/978-981-13-8798-2_7

heart diseases [2]. If a life-threatening disease is not treated timely, it may result in death.

With the advancement of technology and artificial intelligence, it has become possible to predict the diseases in advance and avoid life-threatening situations. There are several machine learning algorithms which can be used to generate disease prediction models [3].

1.1 Machine Learning Algorithms

Broadly, machine learning is classified as supervised learning and unsupervised learning.

1.1.1 Supervised Learning

In this, the model is trained using known class data, that, is both the input and output are known. Based on the differences in the input and predicted output, the connection weights between the neurons are adjusted to minimize the predicted output and actual output. Test data is then fed into the model to check the accuracy of the model. Supervised learning is further classified into regression and classification.

Regression

Regression is a predictive technique in which there is a linear relationship between the dependent (outputs) and independent variable (inputs). Some of the important regression techniques, which can be used for either classification or regression, are simple linear regression, multiple linear regression, polynomial regression, support vector regression, ridge regression, lasso regression, elastic net regression, Bayesian regression, decision tree regression, and random forest regression [4].

Classification

Classification is a noncontinuous prediction technique [5]. In this, the output is not always continuous in nature. Some of the important classification techniques used in machine learning for various applications are logistic regression/classification, K-nearest neighbors, support vector machines, kernel support vector machines, Naïve Bayes, decision tree classification, and random forest classification [6].

Classifiers (models) based on Artificial Neural Network (ANN) also play an important role in the case of classification/prediction. The architecture of ANN consists of three different layers: input layer, hidden layer, and output layer. Each layer has some number of neurons; on the basis of these neurons, information passed through the input layer to the output layer for obtaining the output of prediction/classification [7].

This research article compares K-nearest neighbors, decision tree, support vector machine, Gaussian Naïve Bayes, artificial neural network, and machine learning algorithms.

1.1.2 Unsupervised Learning

Unlike supervised learning where models are trained using known class data, that is, both the input and output are known, unsupervised learning uses datasets without knowledge of class labels.

2 Related Work

To predict the probability of cardiac arrest, ICU transfer, or death, Edelson et al. [8] developed a machine learning analytic gradient boosting machine model. They used only three parameters: age, heart rate, and the respiratory data [8] for training the model.

Hayeri et al. [9] proposed the continuous glucose monitor and the fitness wearable devices for training the algorithms, and the data from each participant was utilized to get significant predictions and accuracy.

Ahmad et al. [10] suggested and discussed heart failure which is a very complex clinical syndrome. This article also shows the various heterogeneous treatments and the responses as per the syndrome to prevent it. In this, the cluster analysis is used and the novel analytic approach can transform the future heart failure clinical trials.

Stewart et al. [11] suggested artificial intelligence-based conceptual techniques that are used to provide the emergency medical support in the recent years; this paper is very important to discuss and give an idea of a research in the field of emergency medicinal situation.

Stone et al. [12] discussed the effects and analysis of calcium formation and effects by which coronary arteries may be blocked due to the calcium formation and as a result of this, various disorders arise. The difficulty in treating as well as diagnosing itself is a challenge. The channel blockage has many bad effects or it can also cause death of human being.

The research article by Churpek et al. [13] discussed the conventional regression for the disease prediction. The research also demonstrates the latest challenges in the said domain of cardiovascular diseases in detail. Garg et al. [4] emphasized the linear regression-based analysis and their impact in predicting the accuracy of liver cancer. The detailed comparison of various algorithms has been done. The study proposes and analyzes the feature subset selection and the usage of the genetic algorithm [14] for the prediction of the heart related disease.

3 Methodology

- **Data Collection**: The data related to cardiovascular disease was collected from the UCI Machine Learning Repository [15]. The models were generated using the Hungarian data. There are 294 instances and 76 attributes but only 14 attributes have been used in the paper. The reason to choose 14 attributes out of 76 is as per the literature survey, only 14 attributes have the relevance in predicting the heart disease. All published experiments have used only 14 attributes out of 76 attributes. The attributes like patients ID number, Social Security Number, pain location, etc., have no relevance in predicting the heart diseases so such types of attributes have been dropped from the dataset.
- **Data Preprocessing**: Some of the instances in the dataset were containing some missing value for one or more attributes, which were replaced by the mean of the attribute. The value of the attribute "num" ranging from 0 to 4 indicates absence (value 0) and presence of heart disease (values 1–4), which has been replaced with attribute 'class' indicating presence (value 1) and absence (value 0) of the disease.
- **Model Generation**: Four different models (classifiers) were generated using the KNN, DT, SVM, GNB, and ANN algorithms. 80% of the total instances were used for training the algorithm.
- **Testing of the Classifiers**: The five of the classifiers generated using the mentioned algorithms in the previous step were tested using the 20% of the total instances and the results (accuracy) obtained recorded.
- **Model Comparison**: The models were compared based on the precision, recall, F1-measures, and accuracy.
- **Results and Discussion**: The testing of the five classifiers has been done using 20% of the total instances (i.e., 59 instances), and the results are analyzed and discussed in the next section.

4 Results and Discussion

Models are compared based on Precision (P), Recall (R), F1-Measure (FM), and accuracy, which can be calculated using the following formulas:

$$P = TP/(TP + FP)$$

$$R = TP/(TP + FN)$$

$$FM = (2 * P * R)/(P + R)$$

$$\text{Accuracy} = (TP + TN)/\text{Total No. of Instances}$$

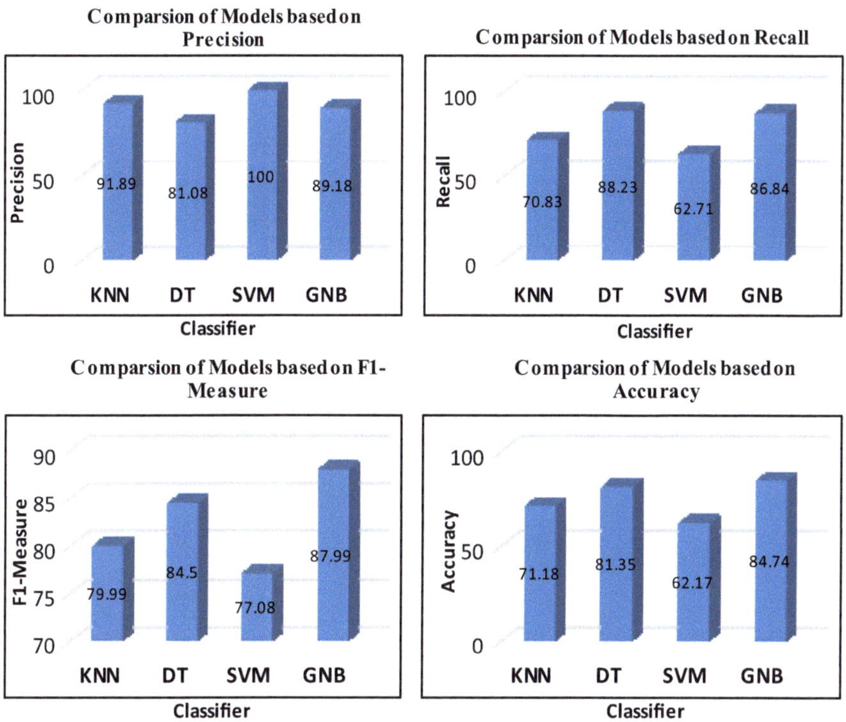

Fig. 1 Comparison of classifiers on various parameters

Table 1 Test results in terms of True Positive (TP), True Negative (TN), False Positive (FP), False Negative (FN), and accuracy. Precision, recall, F1-measure, and accuracy obtained on 59 test instances applied on models generated using algorithms: K-Nearest Neighbors (KNN), decision tree (CART), Support Vector Machine (SVM), and Gaussian Naïve Bayes (GNB)

Classifier (based on algorithm)	TP	TN	FP	FN	P	R	FM	Accuracy (in %)
KNN	34	8	3	14	91.89	70.83	79.99	71.18
CART	30	18	7	4	81.08	88.23	84.50	81.35
SVM	37	0	0	22	100	62.71	77.08	62.71
GNB	33	17	4	5	89.18	86.84	87.99	84.75

where TP stands for True Positive, FP stands for False Positive, FN stands for False Negative, and TN stands for True Negative (Fig. 1).

From Table 1, it is clear that among the four classifiers: KNN, DT, SVM, and GNB, GNB classifier is best in terms of accuracy. Table 2 shows the accuracy observed with ANN as classifier (Fig. 2).

Table 2 Results using artificial neural network

Number of hidden layer	Activation function used	Number of epochs	Accuracy (%)
4	ReLU	100	88.03
4	ReLU	200	91.45

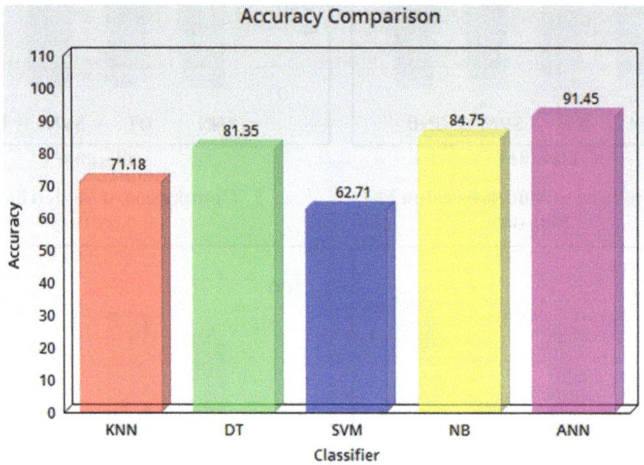

Fig. 2 Comparison of classifiers based on accuracy

5 Conclusion

This paper presents various classifiers such as K-nearest neighbors, decision tree, support vector machine, Gaussian Naïve Bayes, and artificial neural network to predict the heart diseases. It is observed that artificial neural network gives the best results in terms of accuracy compare to all other classifiers, which is 91.45%.

Acknowledgements We acknowledge UCI Machine Learning Repository for providing datasets related to heart diseases. Further, we also acknowledge Hungarian Institute of Cardiology. Budapest: Andras Janosi, M.D., University Hospital, Zurich, Switzerland: William Steinbrunn, M.D., University Hospital, Basel, Switzerland: Matthias Pfisterer, M.D., and V.A. Medical Center, Long Beach and Cleveland Clinic Foundation: Robert Detrano, M.D., Ph.D., for providing the heart diseases-related data.

References

1. https://www.medicalnewstoday.com/articles/237191.php
2. Katagiri, H., Yamada, T., Oka, Y.: Adiposity and cardiovascular disorders: disturbance of the regulatory system consisting of humoral and neuronal signals. Circ. Res. 27–39 (2007)
3. Kieval, R.S.: Drug delivery neural stimulation device for treatment of cardiovascular disorders. US Patent No. 6,178,349. 23 Jan 2001

4. Garg, D., Sharma, A.K.: Prediction and analysis of liver patient data using linear regression technique. In: Advances in Machine Learning and Data Science, pp. 71–80. Springer, Singapore (2018)
5. Schapire, R.E.: The boosting approach to machine learning: an overview. In: Nonlinear estimation and classification, pp. 149–171. Springer, New York (2003)
6. https://towardsdatascience.com/machine-learning-for-beginners-d247a9420dab
7. Ong, M.E.H., et al.: Prediction of cardiac arrest in critically Ill patients presenting to the emergency department using a machine learning score incorporating heart rate variability compared with the modified early warning score. Critical Care (2012)
8. Edelson, D.P., et al.: Less is more: detecting clinical deterioration in the hospital with machine learning using only age, heart rate and respiratory rate. In: Big Data and Artificial Intelligence in Critical Illness. American Thoracic Society (2018)
9. Hayeri, A.: Predicting future glucose fluctuations using machine learning and wearable sensor data. American Diabetes Association (2018)
10. Ahmad, T., et al.: Machine learning methods improve prognostication, identify clinically distinct phenotypes, and detect heterogeneity in response to therapy in a large cohort of heart failure patients. J. Am. Heart Assoc. (2018)
11. Stewart, J., Sprivulis, P., Dwivedi, G.: Artificial intelligence and machine learning in emergency medicine. Emergency Medicine Australasia (2018)
12. Stone, P.H., et al.: Calcium channel blocking agents in the treatment of cardiovascular disorders. Part II: hemodynamic effects and clinical applications. Ann. Intern. Med. 886–904 (1980)
13. Churpek, M.M., et al.: Multicenter comparison of machine learning methods and conventional regression for predicting clinical deterioration on the wards. Crit. Care Med. (2016)
14. Anbarasi, M., Anupriya, E., Iyengar, N.C.S.N.: Enhanced prediction of heart disease with feature subset selection using genetic algorithm. Int. J. Eng. Sci. Technol. 5370–5376 (2010)
15. https://archive.ics.uci.edu/ml/datasets/heart+Disease. Hungarian Institute of Cardiology. Budapest: Andras Janosi, M.D., University Hospital, Zurich, Switzerland: William Steinbrunn, M.D., University Hospital, Basel, Switzerland: Matthias Pfisterer, M.D., V.A. Medical Center, Long Beach and Cleveland Clinic Foundation: Robert Detrano, M.D., Ph.D

A Hybrid Filtering-Based Retinal Blood Vessel Segmentation Algorithm

Piyush Samant, Atul Bansal and Ravinder Agarwal

Abstract The proposed work compares three unsupervised segmentation algorithms for retinal vessel segmentation. In the first stage, uneven light illumination has been removed using morphological operation and with contrast stretching algorithm intensity, normalization has been done. Thereafter, image enhancement has been performed using multiple combinations of filtering algorithms. Finally, using deformable models, Fuzzy c-means and K-means clustering algorithms, segmentation has been performed. After image enhancement using the combination of filters, three segmentation algorithms have been applied and compared. Finally, the segmented images are classified using ground truth standard images and six parameters named as accuracy, sensitivity, specificity, an area under the curve, connectivity area length, and the Matthews Correlation Coefficient have been calculated. All the above-aforementioned algorithms have been tested over three publicly available datasets such as DRIVE, STARE, and CHASE-DB-1.

1 Introduction

In medical diagnostics, information about the health condition of an individual can be predicted by the biomedical images. Different imaging modalities have been used to capture images of different organs and for different purposes. Retinal image analysis is a new technique among all, and is getting popular in ophthalmology as a reliable diagnostic method. A number of eye-related diseases like glaucoma, Diabetic

P. Samant (✉)
Chandigarh University, Gharuan, India
e-mail: piyushsamant88@gmail.com

A. Bansal
GLA University, Mathura, India
e-mail: atulbansal@rediffmail.com

R. Agarwal
Thapar Institute of Engineering and Technology, Patiala, India
e-mail: ravinder_eeed@thapar.edu

M. Gupta et al. (eds.), *Computer Vision and Machine Intelligence in Medical Image Analysis*, Advances in Intelligent Systems and Computing 992,
https://doi.org/10.1007/978-981-13-8798-2_8

Retinopathy (DR), hypertension, etc., can be diagnosed by the retinal images. DR is the most prominent reason for vision loss in the diabetic patients. A total of 285 million people worldwide are affected by the diabetes [1]. Retinal images are acquired from the back of the eye using funds photographic camera, which consists of a flashed camera with a microscopic mechanism [2–4]. To examine the DR from the retinal images, vessels have been segmented and analyzed [5]. Hence, system performance depends on the accurate vessel segmentation, and a wrong segmentation may lead to an inaccurate diagnosis [6].

Vessel segmentation algorithms have been broadly classified into two classes: supervised and unsupervised segmentation algorithms [7, 8]. Supervised algorithms used pre-labeled images in order to train the systems with the help of ophthalmologist [9–14]. Elisa et al. [15] presented an algorithm, which combines the supervised and unsupervised algorithm [16, 17]. Unsupervised segmentation techniques are free from any user interface and involvement. Elena et al. [18, 19] presented an improvement by considering the blood vessels according to their orientation in the retinal image. Maria et al. [20] proposed a morphological and iterative region growing algorithm-based vessel segmentation technique. Al-Diri et al. [21] utilized the ribbon of twins active counter model to capture vessel edge. All the vessel segmentation models discussed above have limitation in segmentation of some very tiny vessels and also the classification models show low sensitivity [22–24]. At the same time, there is a need for fully automatic, fast, and reliable algorithm with no initial configuration of parameters, so that a nonmedical practitioner can also utilize the technology.

2 Proposed Methodology

Figure 1 shows the steps followed and is explained as follows:

2.1 *Nonuniform Illumination Removal and Intensity Normalization*

Nonuniform illumination can occur because of the position of the camera sensor and uneven light reflection. Nonuniformity has been removed using morphological

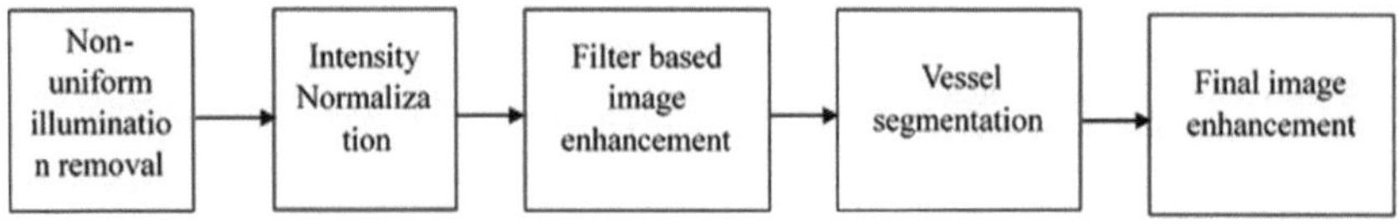

Fig. 1 Diagram of proposed methodology

operations. A conventional top-hat transformation has been applied with the help of multiple construction elements. To normalize the intensity of the images, contrast stretching has been applied to all three planes of the color retinal images [25, 26]. Each pixel of the plane has been normalized in range $(\alpha_1 - \alpha_2)$ as

$$R_0(x) = [R_i(x) - R_i(min)]\left[\frac{\alpha_2 - \alpha_1}{R_i(max) - R_i(min)}\right] + \alpha_1 \tag{1}$$

$$G_0(x) = [G_i(x) - G_i(min)]\left[\frac{\alpha_2 - \alpha_1}{G_i(max) - G_i(min)}\right] + \alpha_1 \tag{2}$$

$$B_0(x) = [B_i(x) - B_i(min)]\left[\frac{\alpha_2 - \alpha_1}{B_i(max) - B_i(min)}\right] + \alpha_1 \tag{3}$$

2.2 *Image Enhancement and Vessel Segmentation*

After contrast normalization of the retinal images, filter-based image enhancement techniques have been applied to enhance the vascular information. Two-dimensional matched filtering [27] techniques give the improved vessel enhancement. Anisotropic-oriented diffusion filter [16] also shows to be the good choice for the small vessels and to remove background noise. In the presented research work, the following combination of filters has been used for further segmentation:

- Anisotropic-oriented diffusion filter followed by Matched filter (MF-AOD),
- Laplacian Gaussian filter followed by Matched filter (MF-LG),
- Anisotropic-oriented diffusion filter followed by Gabor wavelet filters (GW-AOD),
- Laplacian Gaussian filter followed by Gabor wavelet filters (GW-LG).

After image enhancement for every combination of filters, deformable models [28] such as Fuzzy c-means clustering and K-means clustering segmentation algorithms have been applied.

2.3 *Final Post Processing Enhancement*

A final enhancement has been performed to remove small regions those are not elongated structure. First using control limited adaptive histogram equalization (CLAHE) [23] algorithm has been applied.

3 Image Dataset and Evaluation Parameters

Three publically available databases DRIVE [14], STARE [28], and CHASE DB1 [27] are evaluated for the proposed algorithm. These images of different databases have great variability with respec to the light illumination and contrast variation. The error has been calculated as the difference in the segmented image and ground truth images. Accuracy, Sensitivity, Specificity, Area Under the Curve (AUC), Connectivity Area Length (CAL), and Matthews Correlation Coefficient (MCC) parameters have been calculated for the performance evaluation of the classifier.

4 Result and Discussion

Standard ground truth images play a crucial role in evaluating the segmentation performance. Also, the sensitivity and specificity, which have been used as a precision measure in most of the proposed algorithms alone cannot describe the performance of the algorithm. In the present research work, AUC, CAL, and MCC are also computed to evaluate the performance of the classification algorithm. Tables 1, 2, and 3 show the results of all three classification techniques for their combination of different filtering algorithms for DRIVE, STARE, and CHASE DB1 (Fig. 2).

The presented framework produces encouraging results for vessel segmentation using unsupervised algorithms.

Table 1 Performance evaluation for DRIVE dataset

Classification technique	Enhancement using filtering	DRIVE					
		Accuracy (%)	Sensitivity (%)	Specificity (%)	AUC	CAL	MCC
Deformable models	AF-MOD	94.96	81.89	95.25	0.9588	0.5838	0.7892
	MF-LG	95.87	93.21	96.35	0.9614	0.6898	0.7447
	GW-AOD	95.08	80.01	96.38	0.9594	0.6359	0.7744
	GW-LG	95.38	79.10	78.11	0.9294	0.6532	0.7123
Fuzzy C-means	AF-MOD	93.12	81.32	98.16	0.9747	0.6426	0.7387
	MF-LG	94.23	82.34	98.45	0.9118	0.5983	0.7853
	GW-AOD	93.89	74.34	98.02	0.9513	0.6025	0.7736
	GW-LG	94.38	77.61	98.12	0.9327	0.6578	0.7894
K means clustering	AF-MOD	94.73	71.58	98.10	0.8325	0.5871	0.7607
	MF-LG	95.15	75.20	98.06	0.8315	0.7126	0.7023
	GW-AOD	95.23	70.91	77.91	0.8437	0.7014	0.7790
	GW-LG	96.96	81.45	98.66	0.8137	0.7235	0.7870

Table 2 Performance evaluation for STARE dataset

Classification technique	Enhancement using filtering	STARE					
		Accuracy (%)	Sensitivity (%)	Specificity (%)	AUC	CAL	MCC
Deformable models	AF-MOD	95.94	68.69	98.16	0.9253	0.6762	0.7682
	MF-LG	91.44	84.30	92.05	0.2481	0.6815	0.7305
	GW-AOD	89.53	86.06	89.79	0.8784	0.6255	0.7936
	GW-LG	95.74	61.59	99.34	0.8394	0.6727	0.7023
Fuzzy C-means	AF-MOD	95.18	75.38	96.83	0.6834	0.6698	0.7426
	MF-LG	95.71	57.62	98.79	0.8528	0.6122	0.7369
	GW-AOD	95.16	70.24	97.09	0.7020	0.6071	0.7752
	GW-LG	94.67	58.98	98.63	0.7156	0.6515	0.7782
K means clustering	AF-MOD	95.82	69.27	98.39	0.8856	0.7055	0.7175
	MF-LG	95.50	61.79	98.75	0.7506	0.6330	0.7477
	GW-AOD	95.00	74.29	97.02	0.8485	0.6617	0.7433
	GW-LG	94.43	80.46	95.99	0.8065	0.6194	0.7633

Table 3 Performance evaluation for CHASE DB1dataset

Classification technique	Enhancement using filtering	CHASE DB1					
		Accuracy (%)	Sensitivity (%)	Specificity (%)	AUC	CAL	MCC
Deformable models	AF-MOD	94.29	71.18	97.91	0.8397	0.6803	0.7711
	MF-LG	94.17	72.40	97.68	0.8139	0.6193	0.7785
	GW-AOD	94.15	71.03	96.65	0.7935	0.6926	0.7152
	GW-LG	95.36	69.44	98.31	0.7817	0.6306	0.7792
Fuzzy c-means	AF-MOD	95.45	70.27	98.28	0.9102	0.6142	0.7563
	MF-LG	94.95	70.10	97.70	0.8736	0.6201	0.7128
	GW-AOD	95.28	72.38	98.18	0.8536	0.6591	0.7275
	GW-LG	93.13	78.38	97.75	0.8836	0.6438	0.7494
K-means clustering	AF-MOD	94.84	73.07	98.11	0.8968	0.6308	0.7827
	MF-LG	94.89	72.73	98.10	0.7823	0.6821	0.7833
	GW-AOD	94.56	72.42	97.92	0.9025	0.6558	0.7177
	GW-LG	94.48	74.67	98.03	0.8625	0.6520	0.7838

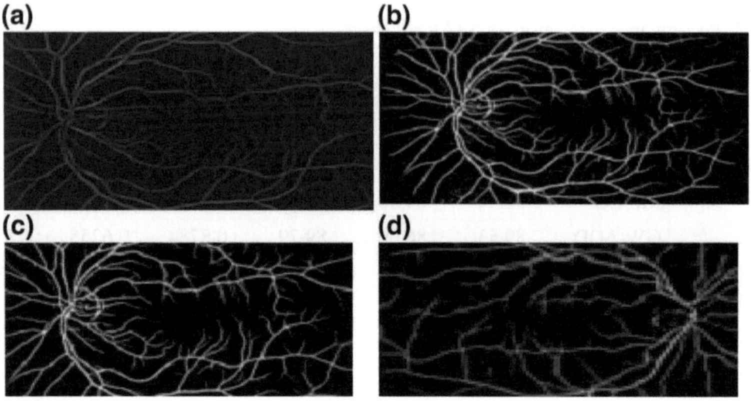

Fig. 2 Filtered result by: **a** Anisotropic-oriented diffusion filter followed by matched filter. **b** Laplacian Gaussian filter followed by matched filter. **c** Anisotropic-oriented diffusion filter followed by Gabor wavelet filters. **d** Laplacian Gaussian filter followed by Gabor wavelet filters

5 Conclusion

In the presented piece of work, a new vessel segmentation algorithm has been developed, which is based on the combination of filters. Comparison of unsupervised segmentation algorithms are also presented for different publicly available databases and different combination of filters used for the image enhancement. The demonstrated performance and effectiveness of the proposed algorithm makes it a suitable choice for vessel segmentation. The proposed algorithm has a great potential for diagnosis application.

References

1. Lee, R., Wong, T.Y., Sabanayagam, C.: Epidemiology of diabetic retinopathy, diabetic macular edema and related vision loss. Eye Vis. **2**(1), 17 (2015)
2. Raju, B., Raju, N.S.D., Akkara, J., Pathengay, A.: Do it yourself smartphone fundus camera—DIYretCAM. Indian J. Ophthalmol. **64**(9), 663 (2016)
3. Gadkari, S.: Innovative model for telemedicine-based screening for diabetic retinopathy in the developing world. Can. J. Ophthalmol. **51**(3), e109–e111 (2016)
4. Rajalakshmi, R., et al.: Validation of smartphone based retinal photography for diabetic retinopathy screening. PLoS ONE **10**(9), 1–10 (2015)
5. Fraz, M.M., et al.: An ensemble classification-based approach applied to retinal blood vessel segmentation. IEEE Trans. Biomed. Eng. **59**(9), 2538–2548 (2012)
6. Pakter, H.M., et al.: Measuring arteriolar-to-venous ratio in retinal photography of patients with hypertension: development and application of a new semi-automated method. Am. J. Hypertens. **18**(3), 417–421 (2005)
7. Niemeijer, M., Staal, J., van Ginneken, B., Loog, M., Abràmoff, M.D.: Comparative study of retinal vessel segmentation methods on a new publicly available database. In: Proceedings of the SPIE 5370, Medical Imaging 2004: Image Processing, pp. 648–656 (2004)

8. Agarwal, D., Bansal, A.: Non-invasive techniques for the screening of diabetic retinopathy. J. Biomed. Imaging Bioeng **1**(2), 1–6 (2017)
9. Huang, Y., Chen, X., Zhang, J., Zeng, D., Zhang, D., Ding, X.: Single-trial ERPs denoising via collaborative filtering on ERPs images. Neurocomputing **149**(PB), 914–923 (2015)
10. Orlando, I., Blaschko, M., Crfs, L.: Learning fully-connected CRFs for blood vessel segmentation in retinal images to cite this version : learning fully-connected CRFs for blood vessel segmentation in retinal images. pp. 634–641 (2014)
11. Lupascu, C.A., Tegolo, D., Trucco, E.: FABC: retinal vessel segmentation using AdaBoost. IEEE Trans. Inf. Technol. Biomed. **14**(5), 1267–1274 (2010)
12. You, X., Peng, Q., Yuan, Y., Cheung, Y.M., Lei, J.: Segmentation of retinal blood vessels using the radial projection and semi-supervised approach. Pattern Recognit. **44**(10–11), 2314–2324 (2011)
13. Marin, D., Aquino, A., Gegundez-Arias, M.E., Bravo, J.M.: A new supervised method for blood vessel segmentation in retinal images by using gray-level and moment invariants-based features. IEEE Trans. Med. Imaging **30**(1), 146–158 (2011)
14. Staal, J., Abràmoff, M.D., Niemeijer, M., Viergever, M.A., Van Ginneken, B.: Ridge-based vessel segmentation in color images of the retina. IEEE Trans. Med. Imaging **23**(4), 501–509 (2004)
15. Ricci, E., Perfetti, R.: Retinal blood vessel segmentation using line operators and support vector classification. IEEE Trans. Med. Imaging **26**(10), 1357–1365 (2007)
16. Soares, J.V.B., Leandro, J.J.G., Cesar, R.M., Jelinek, H.F., Cree, M.J.: Retinal vessel segmentation using the 2-D Gabor wavelet and supervised classification. IEEE Trans. Med. Imaging **25**(9), 1214–1222 (2006)
17. Yin, X., Ng, B.W.H., He, J., Zhang, Y., Abbott, D.: Accurate image analysis of the retina using hessian matrix and binarisation of thresholded entropy with application of texture mapping. PLoS One **9**(4) (2014)
18. Martínez-Pérez, M.E., Hughes, A.D., Stanton, A.V., Thom, S.A., Bharath, A.A., Parker, K.H.: Retinal blood vessel segmentation by means of scale-space analysis and region growing. In: Medical Image Computing and Computer Intervention—MICCAI'99, pp. 90–97 (1999)
19. Martinez-Perez, M.E., Hughes, A.D., Thom, S.A., Bharath, A.A., Parker, K.H.: Segmentation of blood vessels from red-free and fluorescein retinal images. Med. Image Anal. **11**(1), 47–61 (2007)
20. Mendonça, A.M., Campilho, A.: Segmentation of retinal blood vessels by combining the detection of centerlines and morphological reconstruction. IEEE Trans. Med. Imaging **25**(9), 1200–1213 (2006)
21. Al-Diri, B., Hunter, A., Steel, D.: An active contour model for segmenting and measuring retinal vessels. IEEE Trans. Med. Imaging **28**(9), 1488–1497 (2009)
22. Nguyen, U.T.V., Bhuiyan, A., Park, L.A.F., Ramamohanarao, K.: An effective retinal blood vessel segmentation method using multi-scale line detection. Pattern Recognit. **46**(3), 703–715 (2013)
23. Azzopardi, G., Strisciuglio, N., Vento, M., Petkov, N.: Trainable COSFIRE filters for vessel delineation with application to retinal images. Med. Image Anal. **19**(1), 46–57 (2015)
24. Zhao, Y., Rada, L., Chen, K., Harding, S.P., Zheng, Y.: Automated vessel segmentation using infinite perimeter active contour model with hybrid region information with application to retinal images. IEEE Trans. Med. Imaging **34**(9), 1797–1807 (2015)
25. Vermeer, K.A., Vos, F.M., Lemij, H.G., Vossepoel, A.M.: A model based method for retinal blood vessel detection. Comput. Biol. Med. **34**(4), 209–219 (2004)
26. Oliveira, W.S., Teixeira, J.V., Ren, T.I., Cavalcanti, G.D.C.: Unsupervised retinal vessel segmentation using combined filters. PLoS One 1–21 (2016)
27. Chaudhuri, S., Chatterjee, S., Katz, N., Nelson, M., Goldbaum, M.: Detection of blood vessels in retinal images using two-dimensional matched filters. IEEE Trans. Med. Imaging **8**(3), 263–269 (1989)
28. Li, C., Kao, C.-Y.: Minimization of region-scalable fitting energy for image segmentation. IEEE Trans. Image Process **17**(10), 1940–1949 (2009)

Laser Scar Classification in Retinal Fundus Images Using Wavelet Transform and Local Variance

Rashmi Raut, Visharad Sapate, Abhay Rokde, Samiksha Pachade, Prasanna Porwal and Manesh Kokare

Abstract Diabetic retinopathy (DR) affects the vision of the person and may eventually lead to blindness. In the initial stage of the disease, patients are treated with a laser to restrict its progression. Such laser treatment leaves behind scars on the retina and patients are advised to undergo screening regularly to track further complications. This paper presents a novel retinal background characterization approach that explores the potential of discrete wavelet transform and rotational-invariant variance features for texture classification of retinal images with and without laser marks. For this experiment, different classifiers, namely, support vector machine, naive Bayes, neural network, and random forest classifiers are tested. We used two publicly available datasets, namely, LMD-DRS and LMD-BAPT. In all cases, the proposed approach obtained the sensitivity, specificity, and accuracy values higher than 68.9%, 70.2%, and 69.4%, respectively. It was found that all performance measures achieve over 87.5, 89.4, and 86.7% for the classification task using random forest classifier. These promising results suggest that the proposed technique can discriminate retinal images having laser marks and without laser marks, and has the potential to be an important constituent in computerized screening solution for retinal images.

Keywords Retinal image analysis · Laser scar detection · Diabetic retinopathy · Feature extraction · Computer-aided diagnosis · Discrete wavelet transform

1 Introduction

According to International Diabetes Federation (IDF) report [1], all over the world, there were 415 million people suffering from diabetes in 2015. This number may increase up to 642 million people in 2040. Diabetes may lead to complications like diabetic retinopathy (DR), cataract, and glaucoma that may eventually lead to vision

R. Raut (✉) · V. Sapate · A. Rokde · S. Pachade · P. Porwal · M. Kokare
Center of Excellence in Signal and Image Processing, Shri Guru Gobind Singhji Institute of Engineering and Technology, Nanded, India
e-mail: rautrashmi@sggs.ac.in

M. Gupta et al. (eds.), *Computer Vision and Machine Intelligence in Medical Image Analysis*, Advances in Intelligent Systems and Computing 992,
https://doi.org/10.1007/978-981-13-8798-2_9

loss [2]. Hence, people affected with diabetes are advised to do regular screening of the retina. To take care of such a huge population that needs screening of the retina at the regular instances needs a computerized screening solution [3]. Hence, several automated diagnostic systems are developed to detect diseases in retinal images. Usually, the people in advanced stages of DR are treated by laser therapy to stop the progression of vascular leaks [4]. This treatment results in scars on the retina due to high-intensity light beam imposed on the retina. Nowadays, automatic screening of retina plays an important role in prioritizing and timely treatment of the patients [5]. However, the existing computerized frameworks do not allow screening of the patients treated by laser. Hence, there is a need for system that can detect whether the person is already treated by laser or not. It can be done by developing an algorithm that can automatically identify the scars left on the retina by photocoagulation treatment. Detection of these scars may enable the system to automatically differentiate whether a person is already treated or not.

This paper presents a novel approach for the classification of laser and non-laser retinal images using texture descriptors such as discrete wavelet transform (DWT) and rotation-invariant local variance (VAR). The remainder of paper is structured as follows: Sect. 2 presents the literature survey, Sect. 3 details the proposed approach followed by results in Sect. 4 and conclusion in Sect. 5.

2 Related Work

There are very few methods developed for automatic laser scar detection in fundus images. In [6], Syed et al. proposed a system which uses the automatic system for finding the laser marks from colored retinal images. Classification is done using support vector machine (SVM) where it was fed with three color, two texture, and four shape features. The evaluation is done on a locally collected dataset of 380 images with 51 images having laser scars. It gave 94% sensitivity, 97% specificity, and 96% accuracy. In [7], Tahir et al. proposed a system using combination of machine learning and image processing technique. The classification of laser marks was done by minimum distance classifier. The features used consist of nine values calculated by computing color and intensity. The classifiers used in the work were trained and tested over local dataset. Dias et al. [8] proposed a method to spot the laser marks using an automatic detection system. Training dataset had 40 images having laser scars and 176 images without laser scars. A fourfold cross-validation procedure was implemented with 75% of the dataset used for training and 25% of the dataset used for testing. It resulted in sensitivity of 63% and specificity of 99%. Sousa et al. [9] proposed a system which detects automatically the existence of laser scars in retinal images. They used tree-based classifiers. The best result was obtained with simple tree-based classifier with eight features as input. It achieved sensitivity and specificity of 80.0 and 67.6 and 73.3 and 70.6% on publicly available LMD-BAPT dataset using 5 and 50 number of trees, respectively. As discussed very few techniques are developed for laser scar classification and there is a lot of scope for improvement.

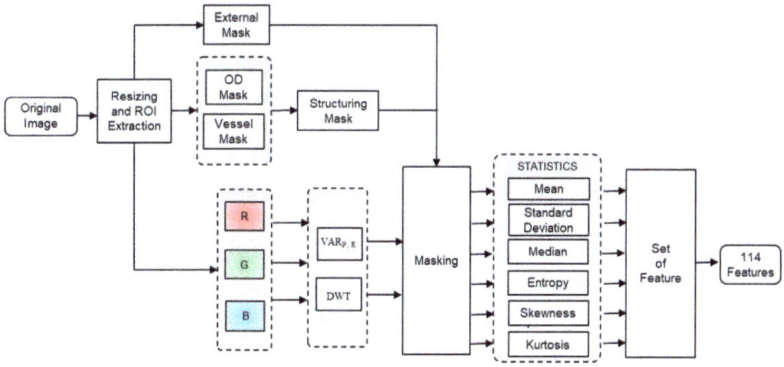

Fig. 1 Block diagram of the proposed method

3 Proposed Method

We propose a novel method in the context of retinal laser mark screening using the textural features. Initially, the input color fundus image (I) is subjected to scale standardization and then region of interest is extracted for further processing. Then each color components, namely, red (I_r), green (I_g), and blue (I_b) are separated and processed to obtain feature set as shown in Fig. 1. For given input image I, discrete wavelet transform (DWT) is applied on I_g and rotational-invariant local variance (VAR) is operated on all three (R, G, and B) components (with neighborhood N = 8 and radius R = 1, 2, 3, 4, and 5). Further, the original image is processed to obtain the optic disk (OD) and vessel mask that are used jointly to generate the structure mask. Additionally, a region of interest (RoI) mask is generated using I_g. All of the masks generated in last step are used for masking the result of the texture descriptors. At last, the final feature set is obtained by computing the statistical measures on image processed with both texture descriptors.

3.1 Preprocessing

This section presents the details of preprocessing steps: image scale standardization and region of interest (RoI) extraction as follows.

3.1.1 Image Scale Standardization

Retinal image standardization is an essential step before extracting textural information. Images used in this study are of different sizes. Further, the VAR measure varies with the change in the radius of neighborhood. Hence, the image scale is standardized with reference to a fixed resolution using following expression:

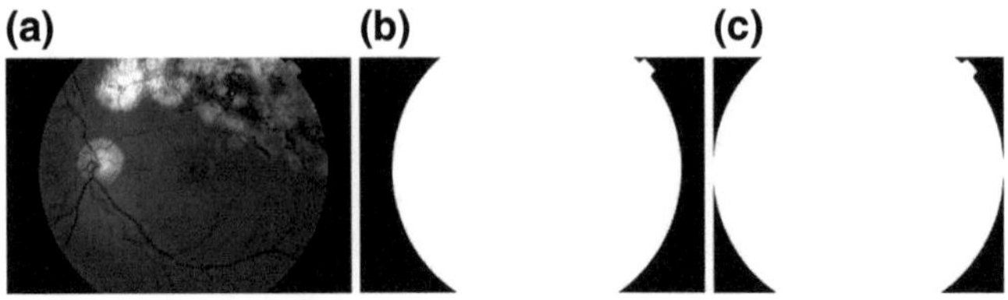

Fig. 2 Illustration of (**a**) fundus image I_g, (**b**) its FOV, and (**c**) RoI

$$R_e = \alpha \left[\frac{A_{px}(I)}{A^{\theta}_{\mathrm{mm}^2}(Retina)} \right]^{1/2} \tag{1}$$

where $A_{px}(I)$ denotes the area (in pixels) of input image and $A^{\theta}_{\mathrm{mm}^2}(Retina)$ is the area of captured retina (in mm^2). α is the scaling factor for transformation of the image I to the chosen resolution R_e. $A^{\theta}_{\mathrm{mm}^2}$ depends on the field of view (FOV) of the input image and the corresponding value of $A^{\theta}_{\mathrm{mm}^2}$ for 45° FOV is 124.8 mm^2 [10]. We choose resolution $R_e = 50$ pixels/mm to standardize all images.

3.1.2 Region of Interest Extraction

Once the image is resized, region of interest (RoI) is extracted using green plane (see Fig. 2a) of the color fundus image I defined as $I_g = (I_g(i))_{i \in \mathbb{R}^2}$. I_g is chosen over I_r and I_b planes as it has better contrast between the retinal atlas and the background. RoI extraction is important step required for two reasons: (a) To reduce the region being processed in the further steps and (b) To remove the texture information generated by the edges of (FOV) mask in the last step. The FOV as shown in Fig. 2b is determined as follows:

$$FOV = \begin{cases} 1, & g(i) \geq g(mean)/2 \\ 0, & \text{otherwise} \end{cases} \tag{2}$$

Now the RoI (as shown in Fig. 2c) is estimated by using the region properties such as major and minor axis lengths to crop the boundaries of the FOV mask.

3.2 *Structural Mask Generation*

Optic disk and vessels are the normal retinal structures. It is important to detect and remove these regions to avoid any interference to the textural information generated using the DWT and VAR. This section presents the approach used for segmentation of optic disk and vessels that are later used for masking the results of texture descriptors as shown in Fig. 1.

3.2.1 Optic Disk Segmentation

The optic disk segmentation is done using L_0 gradient ($Grad_{L_0}$) minimization based smoothing approach [11]. The $Grad_{L_0}$ evaluates the degree of flatness by counting the number of pixels whose horizontal (h) and vertical (v) difference is nonzero. Let $I_{OD} \in R^{h \times v}$ is input image and S represents smoothened image. The $Grad_{L_0}$ of S is given as

$$Grad_{L_0}(S) = \sum_{x=1}^{h}\sum_{y=1}^{v} F\Big(|S_{x+1,y} - S_{x,y}| + |S_{x,y+1} - S_{x,y}|\Big) \tag{3}$$

where $\partial_h S_P = |S_{x+1,y} - S_{x,y}|$ and $\partial_v S_P = |S_{x,y+1} - S_{x,y}|$. The objective function of smoothing by L_0 gradient minimization is defined as follows:

$$f = \min_{S}\Big\{\sum_{p}(S_p - I_{OD_p})^2 + \lambda.Grad_{L_0}(S)\Big\} \tag{4}$$

where λ is the parameter controlling significance of $Grad_{L_0}(S)$, and in our work we use $\lambda = 0.001$. The smoothing function is solved using the alternating optimization strategy by inducting auxiliary variables. The smoothened image is segmented using the Otsu's thresholding algorithm and further processed by morphological dilation using disk structure of size 5. The final segmented OD is as shown in Fig. 3c.

3.2.2 Vessel Segmentation

We used our earlier unsupervised technique [12] to segment the blood vessels from the retina. Initially, vessel enhancement is done using an adaptive manifold filter. In this filter weighted average of I_g is computed and an enhanced image E_{I_g} is formed by a new set of pixels. Image is further enhanced using contrast limited adaptive histogram equalization (CLAHE) and then 2-D median filtering is performed on the image. Finally, mathematical morphology and iterative thresholding operations are done to segment the blood vessels (as shown in Fig. 3b). Now both OD and vessels mask are combined to obtain the joint structure mask as shown in Fig. 3d.

3.3 *Feature Extraction*

The feature extraction from the retinal images is done using discrete wavelet transform and local variance operator as follows.

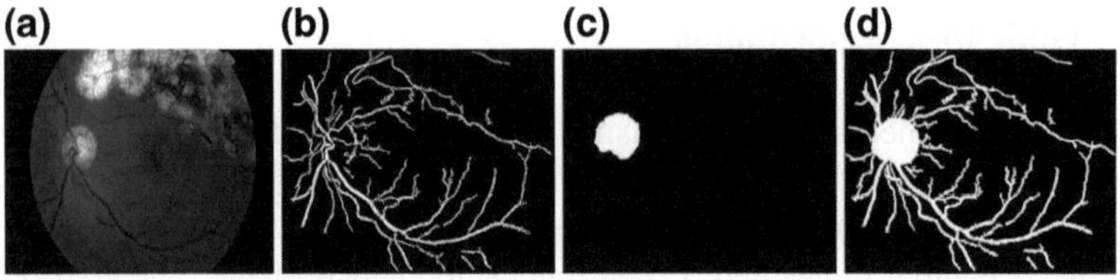

Fig. 3 Generation of Structure Mask: **a** Input image I_g. **b** Segmented vessels. **c** Segmented OD, and (**e**) Final joint structure mask

3.3.1 Discrete Wavelet Transform

To take advantage of both spatial and transform domain information in the retinal images, we employ DWT [13] to compute the features. The DWT decomposes a signal into basis functions which are obtained by performing shifting (translation) and the scaling (dilation) operations on a single wavelet function called as mother wavelet. The first level of detailed approximate coefficients $Y_{HPF}(k)$ and $Y_{LPF}(k)$ of signal is obtained by passing the input signal $I_g(n)$ through half band of high-pass $h(2k-n)$ filter and low-pass filter $l(2k-n)$ separately. They are represented as

$$Y_{HPF}(k) = \sum_{n=-\infty}^{\infty} I_g(n)h(2k-n), Y_{LPF}(k) = \sum_{n=-\infty}^{\infty} I_g(n)l(2k-n) \tag{5}$$

where the mother wavelet is defined as follows:

$$\psi_{m,n}(x) = 2^{\frac{-m}{2}}\psi(2^{-m}x-n) \tag{6}$$

We have employed Haar wavelet with one level of decomposition for generation of feature images. The resultant DWT feature images after removal of structure mask are shown in Fig. 4b–e.

3.3.2 Variance

We have computed the rotational-invariant local variance (*VAR*) [14] that is effective in generating the local contrast features represented as follows:

$$VAR_{N,R} = \frac{1}{N}\sum_{n=0}^{N-1}(G_N-\mu)^2, \mu = \frac{1}{N}\sum_{n=0}^{N-1}G_N \tag{7}$$

where N is the number of points and G_N is the intensity value of its neighborhood. The VAR is calculated for every pixel having neighborhood defined by the input arguments. The VAR operator computes variance on the circumference of a circle with the radius R and the circumference is discretized into N equally spaced sample

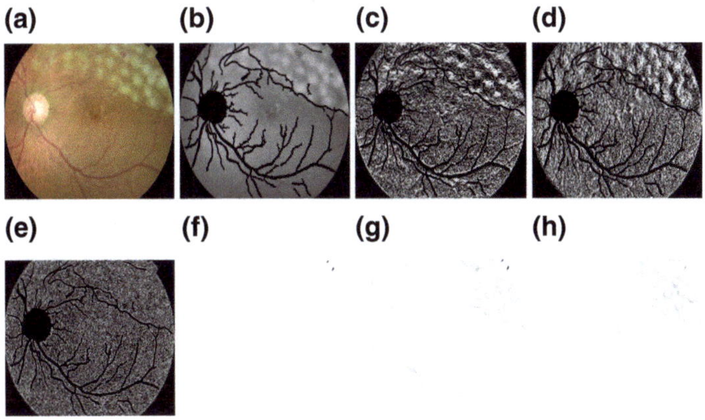

Fig. 4 Generation of feature images: **a** Input image I, (**b–e**) DWT feature images for LL, LH, HL, and HH bands, respectively, and (f-h) VAR feature images for I_r, I_g, and I_b planes (results shown for $N = 8$ and $R = 5$)

points. For every pixel of RGB image, VAR is measured using $N = 8$ and R ($R = 1, 2, 3, 4, 5$).

Finally, the VAR and DWT measures are united to boost the performance and find features. It is important to note that we have individually analyzed all color planes for generation of VAR features unlike for the wavelet where we computed features on green plane only. After applying the structure and external mask (as shown in Fig. 4) the image values are collected into histograms. For each color plane, the VAR images are used to compute the histograms. Similarly, for the DWT features generated using green plane all values are collected into the histogram. Now, these generated histograms are utilized to obtain different statistical features. We compute six statistical features, namely, mean, entropy, standard deviation, kurtosis, skewness, and median. In case of VAR, the total number of features are generated by 6 statistical features $\times$ 5 radius $\times$ 3 components that sum up to 90. Similarly, six statistical features are generated from DWT for four wavelet coefficients on green component that equals to 24. So, the total number of features extracted are 114.

3.4 *Classification*

After the extraction of features, the classification approach [15] is utilized in which the training set is preprocessed before the classification step. The preprocessing employs two steps: (a) data normalization and (b) data resampling. Data normalization is important because the values of data vary widely. It helps in standardizing attributes in the data at zero mean and unit variance. Whereas data resampling is done because the data in hand is highly unbalanced that negatively affects the performance of the machine learning algorithm. The synthetic minority oversampling

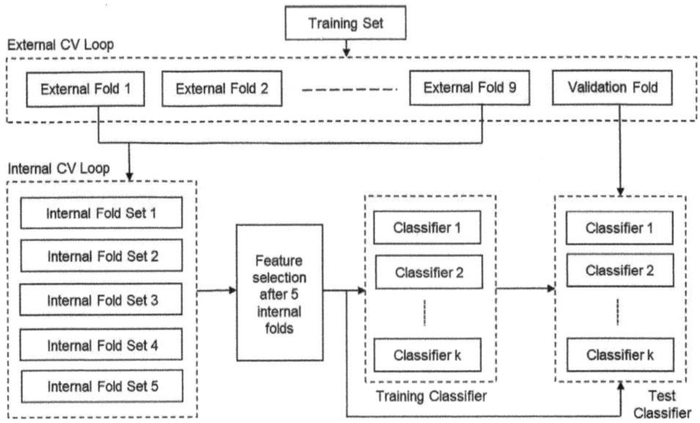

Fig. 5 External cross-validation procedure

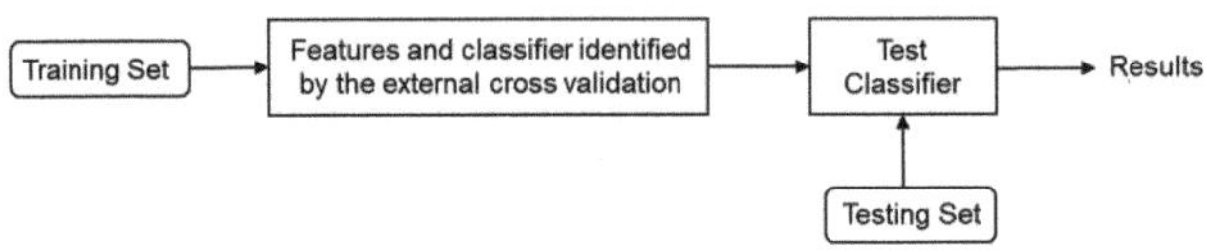

Fig. 6 Final cross-validation procedure

technique (SMOTE) is used for resampling to reduce such negative effects. Then nested cross-validation (CV) also called external CV (as shown in Fig. 5) is executed on the training set to reduce the dimensionality of data by means of feature selection before sending to the classifier (Fig. 6).

Feature selection is done by forward (best first search) approach. Tenfolds are used in the external loop and fivefolds in the internal loop. The external loop and internal loop use the same classifier type. The external loop divides the set into 10 nonoverlapping pairs such that it constitutes 90% images for training and 10% for testing. The nested CV gives a measure of the performance of the proposed approach on dataset used for experimentation. Classification and data preprocessing were performed in Weka.

4 Results

We used two publicly available datasets, namely, laser marks dataset—diabetic retinopathy screening (LMD-DRS) and laser marks dataset—before and after photocoagulation treatment (LMD-BAPT) contained total of 49 images out of which 34 has laser marks and 15 without laser marks. LMD-DRS contained 622 total images out of which 203 has laser marks and 419 without laser marks [9].

We used naive Bayes, neural network, random forest, and SVM as classifier with a different number of trees, i.e., 5 and 50 for random forest. Measures like sensitivity (SE), specificity (SP), and overall accuracy (ACC) give the performance of the proposed system as given in Tables 1 and 2.

Table 1 Classifier comparison on training and testing set

LMD-DRS				LMD-BAPT			
Experiments	SE	SP	ACC	Experiments	SE	SP	ACC
External CV on training set				External CV on training set			
Naive bayes	75.1	84.4	79.8	Naive bayes	87.4	89.5	88.4
Neural network	81.8	82.7	81.9	Neural network	88.9	95.8	93.1
Random forest	**92.5**	**93.4**	**93.0**	Random forest	**94.5**	**97.4**	**95.2**
SVM	78.3	77.4	77.9	SVM	82.4	91.4	85.9
Test set				Test set			
Naive bayes	68.9	70.2	69.4	Naive bayes	75.7	73.5	74.9
Neural network	78.6	85.7	83.2	Neural network	86.7	87.5	87.3
Random forest	**87.5**	**89.4**	**86.7**	Random forest	**88.9**	**95.8**	**92.7**
SVM	76.0	79.4	77.8	SVM	79.1	86.4	84.4

Table 2 Comparison of proposed method with existing approach on LMD-BAPT dataset

RF 5				RF 50			
Methods	SE	SP	ACC	Methods	SE	SP	ACC
Sousa et al. [9]	80.0	67.6	–	Sousa et al.	73.3	70.6	–
Proposed	**87.5**	**89.4**	**86.7**	Proposed	**88.9**	**95.8**	**92.7**

5 Conclusion

In this paper, we proposed a novel method to characterize retinal fundus images for discriminating between laser treated and non-treated images. The novelty of this work is that it does not require prior segmentation of the laser scars for classification purpose. We tested our approach using four different classifiers that suggest random forest is well suitable for the purpose. The encouraging results obtained by the information encoded using DWT and VAR highlight its suitability as an integrable solution for screening of diabetic retinopathy.

References

1. International Diabetes Federation. Diabetes: facts and figures (2015)
2. Sayin, N., Kara, N., Pekel, G.: Ocular complications of diabetes mellitus. World J. Diabetes **6**(1), 92 (2015)
3. Jordan, K.C., Menolotto, M., Bolster, N.M., Livingstone, I.A., Giardini, M.E.: A review of feature-based retinal image analysis. Expert Rev. Ophthalmol. **12**(3), 207–220 (2017)
4. Zhang, W., Liu, H., Rojas, M., Caldwell, R.W., Caldwell, R.B.: Anti-inflammatory therapy for diabetic retinopathy. Immunotherapy **3**(5), 609–628 (2011)

5. Porwal, P., Pachade, S., Kokare, M., Deshmukh, G., Sahasrabuddhe, V.: Automatic retinal image analysis for the detection of diabetic retinopathy. In: Biomedical Signal and Image Processing in Patient Care, pp. 146–161. IGI Global (2018)
6. Syed, A.M., Akbar, M.U., Akram, M.U., Fatima, J.: Automated laser mark segmentation from colored retinal images. In: 2014 IEEE 17th International Multi-Topic Conference (INMIC), pp. 282–286. IEEE (2014)
7. Tahir, F., Akram, M.U., Abbass, M., Khan, A.A.: Laser marks detection from fundus images. In: 2014 14th International Conference on Hybrid Intelligent Systems (HIS), pp. 147–151. IEEE (2014)
8. Dias, J.M.P., Oliveira, C.M., da Silva Cruz, L.A.: Detection of laser marks in retinal images. In: 2013 IEEE 26th International Symposium on Computer-Based Medical Systems (CBMS), pp. 532–533. IEEE (2013)
9. e Sousa, J.G.R.A., Oliveira, C.M., da Silva Cruz, L.A.: Automatic detection of laser marks in retinal digital fundus images. In: 2016 24th European Signal Processing Conference (EUSIPCO), pp. 1313–1317. IEEE (2016)
10. Molina-Casado, J.M., Carmona, E.J., García-Feijoó, J.: Fast detection of the main anatomical structures in digital retinal images based on intra-and inter-structure relational knowledge. Comput. Methods Programs Biomed. **149**, 55–68 (2017)
11. Porwal, P., Pachade, S., Kadethankar, A., Joshi, A., Patwardhan, V., Kamble, R., Kokare, M., Meriaudeau, F.: Automatic segmentation of optic disc by gradient minimization based approach. In: 2018 International Conference on Intelligent and Advanced System (ICIAS), pp. 1–5. IEEE (2018)
12. Pachade, S., Porwal, P., Kokare, M.: A novel unsupervised framework for retinal vasculature segmentation. In: Advanced Computational and Communication Paradigms, pages 490–497. Springer (2018)
13. Starck, J.-L., Murtagh, F.D., Bijaoui, A.: Image Processing and Data Analysis: the Multiscale Approach. Cambridge University Press (1998)
14. Ojala, T., Pietikäinen, M., Mäenpää, T.: Gray scale and rotation invariant texture classification with local binary patterns. In: European Conference on Computer Vision, pp. 404–420. Springer (2000)
15. Porwal, P., Pachade, S., Kokare, M., Giancardo, L., Mériaudeau, Fabrice: Retinal image analysis for disease screening through local tetra patterns. Comput. Biol. Med. **102**, 200–210 (2018)

Automated Segmentation of Cervical Cells Using MSER Algorithm and Gradient Embedded Cost Function-Based Level-Set Method

Kaushiki Roy, Debotosh Bhattacharjee and Mita Nasipuri

Abstract Traditionally, cervical cells are screened by analyzing Pap smear slides. But this manual inspection requires expert pathologist making the entire process time consuming and prone to manual errors. Thus, it is needed urgently to develop an automated system for the screening process. Though extensive research work is going on for decades to develop the automated system, but the success is quite less owing to the fact in Pap smears, nuclei and cytoplasm are often found in clumps lacking any boundaries separating them. In this work, we have proposed a gradient embedded cost function for cytoplasm segmentation. We have used ISBI-15 dataset for the work and the result obtained is compared with the state-of-the-art techniques.

Keywords Cervical cell · MSER algorithm · Pap smear · Level set · C-V model · Gradient image

1 Introduction

According to reports [1] published in 2018, cervical cancer is the fourth major cause of cancer amongst females worldwide. In [2], superpixel, triangle thresholding [3], and graph cuts were used for cervical cell segmentation. In [4], global and local graph cuts-based approach were incorporated for cytoplasm and nucleus segmentation. In [5], thresholding, binary classifier was used for cervical cell segmentation. In [6], k-means clustering and radiating gradient vector flow (RGVF) snake algorithm were used for cervical nuclei and cytoplasm segmentation. In [7], level-set-based approach with shape prior was used to detect cell nuclei in cervical images. In [8,

K. Roy (✉) · D. Bhattacharjee · M. Nasipuri
Computer Science and Engineering Department, Jadavpur University, Kolkata, India
e-mail: kaushiki.cse@gmail.com

D. Bhattacharjee
e-mail: debotosh@cse.jdvu.ac.in

M. Nasipuri
e-mail: mnasipuri@cse.jdvu.ac.in

M. Gupta et al. (eds.), *Computer Vision and Machine Intelligence in Medical Image Analysis*, Advances in Intelligent Systems and Computing 992,
https://doi.org/10.1007/978-981-13-8798-2_10

9] level-set based approach was used for cervical segmentation. Song et al. [10] have proposed a deep learning-based framework for cervical nuclei and cytoplasm segmentation. However, segmenting nuclei and cytoplasm in highly overlapping cell clusters remains a big challenge till date. This research paper is focused at developing an automated framework for overlapping nuclei and cytoplasm segmentation using MSER algorithm and gradient embedded cost function (GCF)-based-level-set algorithm.

2 Methodology

The overview of the proposed framework is shown in Fig. 1.

2.1 Clump Detection

This is the first phase of our framework, which used C-V [11] model for segmentation of isolated as well as cellular clumps. The model works appropriately well even in cases where boundaries are irregular possessing no fixed boundary pattern.

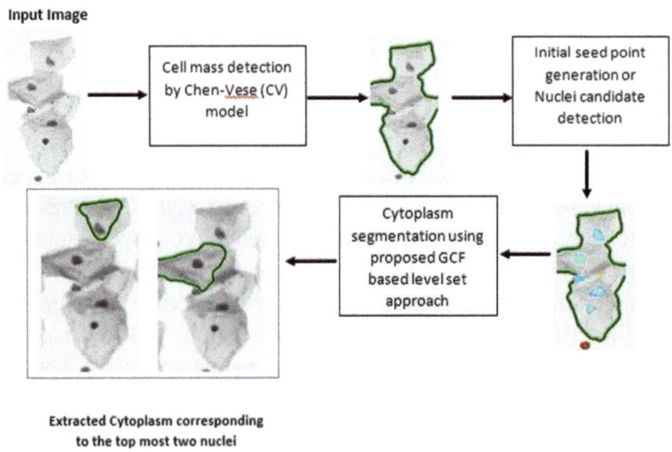

Fig. 1 Block diagram for segmentation of nuclei and cytoplasm from overlapping cervical cells

2.1.1 Mathematical Formulation of Level-Set Algorithm Using C-V Model

A 2-D cervical image $\omega_0(x, y)$ consist of background and foreground represented by B_c and B_f, respectively. Curve C separates these two regions. Let τ be the implicit function, which is used to represent the separating curve C. Further, we assume that the image domain is represented by D. Assuming φ^+ represent the outside region that is the background region, whereas φ^- represent foreground or the cellular region. The value of τ for cellular regions (φ^-) is <0 that is in the cellular region $\tau < 0$. In contrary, $\tau > 0$, for background region (φ^+) and in boundary, $\tau = 0$. Consider θ_1 as the mean intensity of foreground region and θ_2 the mean intensity of background. Equation 1 is the objective function or the energy functional to be minimized by the evolving level-set curve.

$$\begin{aligned} E(C) = \lambda_1 \int_{\varphi} |\omega_0(x, y) - \theta_1|^2 H(\tau(\mathrm{x}, \mathrm{y}))dxdy \\ + \lambda_2 \int_{\varphi} |\omega_0(x, y) - \theta_2|^2 \big[1 - H(\tau(\mathrm{x}, \mathrm{y}))\big]dxdy \\ + \mu \int_{\varphi} \sigma(\tau(\mathrm{x}, \mathrm{y}))|\nabla(\tau(\mathrm{x}, \mathrm{y})| + v \int_{\varphi} H(\tau(\mathrm{x}, \mathrm{y}))dxdy \end{aligned} \tag{1}$$

The parameters λ_1, λ_2, μ, and v are fixed parameters and their value is decided by the user. In our work, we used $\lambda_1 = 1$, $\lambda_1 = 1$, $\mu = 1$, and $v = 0$. The Heaviside function $H(\tau)$ is represented by Eq. 2.

$$H(\tau) = \begin{cases} 0 \; if \;\; \tau \le 0 \; that \, is \, inside \, the \, region \; (cell) or \; on \; the \; boundary \\ 1 \; if \;\; \tau > 0 \; that \, is \, outside \, the \, region \; namely \; the \; background \end{cases} \tag{2}$$

The derivative of Heaviside function $H(\tau)$ is called the Dirac delta function represented by $\sigma(\tau)$ that is

$$\sigma(\tau) = H'(\tau) \tag{3}$$

The delta function ($\sigma(\tau)$) has positive value (=1) only on the boundary that is $\tau = 0$, rest for all the others ($\tau < 0 \; or \; \tau > 0$), its value is zero. Thus, proper approximation to this integral is not provided by any standard approximation algorithm, which forces us to use a first-order smeared out approximation of $\sigma(\tau)$. This, in turn, necessitates re-defining a smeared out Heaviside function as

$$H(\tau) = \begin{cases} 0 & \tau < -\varepsilon \\ \frac{1}{2} + \frac{\tau}{2 \in} + \frac{1}{2\pi} \mathrm{Sin}\left(\frac{\pi \tau}{\in}\right) & -\varepsilon \le \tau \le \varepsilon \\ 1 & \varepsilon < \tau \end{cases} \tag{4}$$

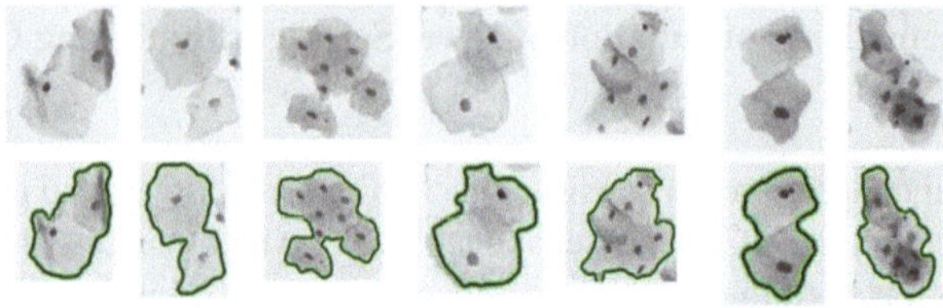

Fig. 2 The top row denotes the free-lying and overlapping cervical cells and the bottom row indicates their extracted initial contours

Likewise, the Dirac delta function can be redefined as shown in Eq. 5.

$$\vartheta(\tau) = \begin{cases} 0 & \tau < -\varepsilon \\ \dfrac{1}{2\in} + \dfrac{1}{2\in}\cos\left(\dfrac{\pi\tau}{\in}\right) & -\varepsilon \le \tau \le \varepsilon \\ 1 & \varepsilon < \tau \end{cases} \tag{5}$$

Next, we differentiate the energy functional defined in Eq. 1 with respect to τ and get

$$\frac{\partial E}{\partial \tau} = \lambda_1 |\omega_0(x, y) - \theta_1|^2 - \lambda_2 |\omega_0(x, y) - \theta_2|^2 \tag{6}$$

The energy functional is minimized by gradient descent flow. Equation for the same is presented in 7.

The overall energy functional is minimized by gradient descent flow as shown in Eq. 9.

$$\frac{\partial \tau}{\partial t} = -\frac{\partial E}{\partial \tau} \tag{7}$$

Equation 8 represents the update equation of τ.

$$\tau^{n+1} = \tau^n + \Delta t \frac{\partial \tau}{\partial t} = \tau^n - \Delta t \frac{\partial \tau}{\partial t} \tag{8}$$

The present iteration is represented by τ^n, whereas the next iteration by τ^{n+1}. The change in time or timestamp is represented by Δt. The underlying figure (Fig. 2) shows cell mass detection by C-V model for both isolated and cellular clumps. These initial rough contours of cell masses are passed to the next phase for nuclei detection.

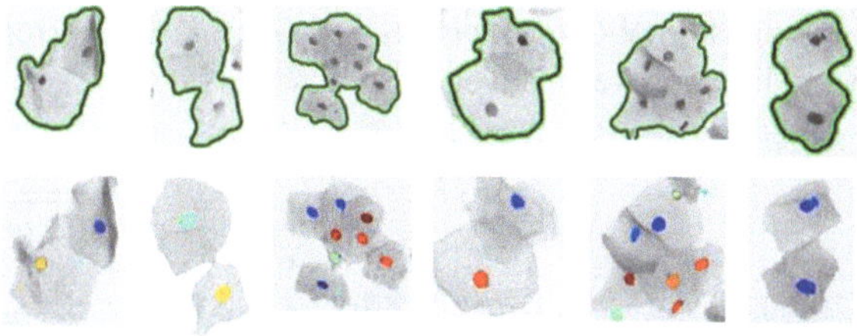

Fig. 3 The top row includes images obtained after clump detection in step 2.1 and the bottom row indicates the potential nuclei candidates

2.2 Nuclei Extraction

This is the second phase of our proposed framework. As evident from Fig. 2, in a cell, nucleus is much darker than cytoplasm, possessing roughly circular feature. These factors allowed us to exploit the potentiality of maximally stable extremal regions (MSER) [12] algorithm for nuclei detection since MSER detect those blobs that possess higher intensities than their surroundings. These higher intensity blobs are the nuclei candidates in our work. Some higher intensity non-nuclei pixels were detected as well by MSER algorithm but we successfully eliminated them by circularity check. Nuclei detected from cell masses by the proposed algorithm is shown in Fig. 3.

2.3 Cytoplasm Extraction

This is the most significant phase of our work that deals with overlapping cytoplasm segmentation. The nuclei of each cell detected in the previous phase is passed to this phase to serve as the initial seed point for the level-set curve to start evolving. Traditional C-V model suffers from one major drawback since it assumes an image to have constant piece-wise intensity. As a result, though the traditional C-V model segments the cell mass from the background but fails to detect the internal boundaries separating one cell from the other in cell chunks since the overlapping cell regions are not made of constant piece-wise intensities, which is clearly depicted in Fig. 2. But the gradient values of these overlapping regions are very high owing to the fact multiple cells overlap there making it a very high-intensity region. We took advantage of this phenomenon and in our proposed gradient embedded cost function, the gradient information is added to the energy functional of the traditional C-V model.

2.3.1 Mathematical Formulation of GCF-Based Level-Set Segmentation

The 2D gradient image for the input cervical image $\omega_0(x, y)$ is represented by $Grad_x$ and $Grad_y$, respectively. These gradient images for an input image is calculated using

Gaussian kernel $GausKernel_1$ and $GausKernel_2$ (represented in Eqs. 9 and 10) in X–Y direction, respectively.

$$GausKernel_1(x, y, \omega) = \frac{-x}{\sqrt{2\pi\omega^3}} e^{-\frac{x^2+y^2}{2\omega^2}} \tag{9}$$

$$GausKernel_2(x, y, \omega) = \frac{-y}{\sqrt{2\pi\omega^3}} e^{-\frac{x^2+y^2}{2\omega^2}} \tag{10}$$

Here, ω represents the width of the Gaussian kernel, whereas ω_0 represents the image function.

The modified energy functional to be minimized, which incorporates the gradient information is given in Eq. 11. Most of the notations used in this section are the same as in Sect. 2.1.

$$\mathrm{E}(total) = E(C) + E_{grad}(x, y, \omega) \tag{11}$$

$$\begin{aligned} E_{grad}(x, y, \omega) = &\int_{\varphi} GausKernel_1(x, y, \omega) * (|\omega_0(x, y) - \theta_1|)^2 H(\tau(\mathrm{x}, \mathrm{y}))dxdy \\ &+ \int_{\varphi} GausKernel_2(x, y, \omega) \\ &* (|\omega_0(x, y) - \theta_2|)^2 \big([1 - H(\tau(\mathrm{x}, \mathrm{y}))\big]dxdy\big) \end{aligned} \tag{12}$$

The $*$ represents the convolution operator. Finally, by elaborating Eq. 11, we get Eq. 13.

$$\begin{aligned} \mathrm{E}(total) = &\lambda_1 \int_{\varphi} |\omega_0(x, y) - \theta_1|^2 H(\tau(\mathrm{x}, \mathrm{y}))dxdy + \lambda_2 \int_{\varphi} |\omega_0(x, y) \\ &- \theta_2|^2[1 - H(\tau(\mathrm{x}, \mathrm{y}))]dxdy + \mu \int_{\varphi} \sigma(\tau(\mathrm{x}, \mathrm{y}))|\nabla(\tau(\mathrm{x}, \mathrm{y})| \\ &+ v \int_{\varphi} H(\tau(\mathrm{x}, \mathrm{y}))dxdy + \int_{\alpha} GausKernel_1(x, y, \omega) * (|\omega_0(x, y) \\ &- \theta_1|)^2 H(\tau(\mathrm{x}, \mathrm{y}))dxdy + \int_{\alpha} Gaus\mathrm{Kernel}_2(x, y, \omega) \\ &* (|\omega_0(x, y) - \theta_2|)^2([1 - H(\tau(\mathrm{x}, \mathrm{y}))]dxdy) \end{aligned} \tag{13}$$

On differentiating Eq. 13 with respect to τ, we get 14.

$$\begin{aligned} \frac{\partial E}{\partial \tau} = &\lambda_1|\omega_0(x, y) - \theta_1|^2 - \lambda_2|\omega_0(x, y) - \theta_2|^2 \\ &+ GausKernel_1(x, y, \omega) * (|\omega_0(x, y) - \theta_1|)^2 \\ &+ GausKernel_2(x, y, \omega) * (|\omega_0(x, y) - \theta_2|)^2 \end{aligned} \tag{14}$$

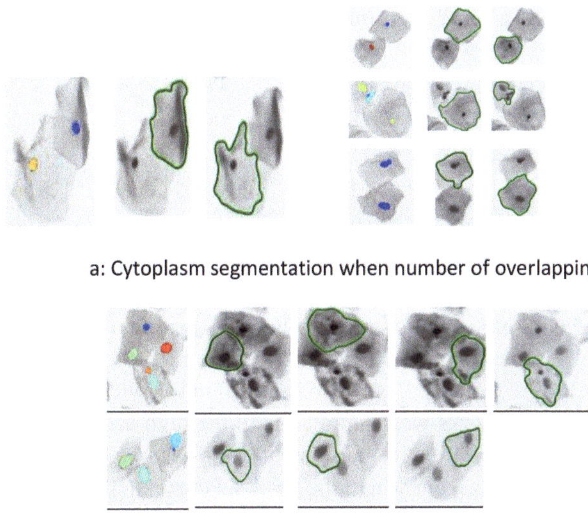

a: Cytoplasm segmentation when number of overlapping cells is 2.

b: Cytoplasm segmentation when number of overlapping cells is greater than 2.

Fig. 4 Cytoplasm segmentation results generated by our proposed algorithm using each detected nuclei as initial seed points when (**a**) number of overlapping cells is 2 and (**b**) number of overlapping cells is greater than 2

The gradient descent flow used for minimizing energy functional and the update equation is the same as Eqs. 7 and 8. Figure 4 shows the results of the cytoplasm extraction obtained through this approach.

3 Experimental Results

3.1 Database Used

ISBI-2015 [13] challenge dataset has been used for this work and this dataset consist of 16 real cervical extended depth of field images and 945 synthetic images.

3.2 Evaluation Metrics

Dice coefficient (DC), True Positive Rate (TP_r), and False Negative Rate (FN_r) have been used to evaluate the performance of our proposed framework. Table 1 shows the comparison of our proposed technique with some other techniques mentioned in the state of the art and the results obtained are highly satisfactory.

Table 1 Comparative evaluation of cytoplasmic segmentation obtained using our proposed algorithms and other works in the state of the art

Dataset	Method	DC	(TP_r)	(FN_r)
ISBI-15	Lee et al. [1]	0.897	0.882	0.137
	Nosrati et al. [9]	0.871	0.875	0.111
	Lu et al. [10]	0.893	0.905	0.316
	Our approach	0.970	0.905	0.094

The ISBI-2015 dataset evaluation metrics states algorithm, which achieves DC > 0.7 for cytoplasm segmentation is considered to be a good algorithm. This framework achieves DC ~ 0.970 and is thus, a good algorithm for cytoplasm segmentation. The proposed algorithm works well mainly due to the high accuracy of nuclei segmentation phase (phase 2), which further increases the accuracy of phase 3 since the nuclei serve as initial points from which the level-set curve starts to evolve. Second, the proposed noble gradient embedded energy functional increases the accuracy of cytoplasm segmentation phase and works well even when there are large number of overlapping cells in the clumps.

4 Conclusion and Future Work

In this work, we have used gradient embedded cost function to segment overlapping cervical cells. The dataset used for this work is ISBI-2015, which consist of large number of cellular clusters. This work performs well mainly due to the exact selection of initial seed points and the proposed energy functional for cytoplasm segmentation.

Acknowledgements Miss Kaushiki Roy is an Inspire fellowship awardee (IF170366).

References

1. Bray, F., Ferlay, J., Soerjomataram, I., Siegel, R.L., Torre, L.A., Jemal, A.: Global cancer statistics 2018: GLOBOCAN estimates of incidence and mortality worldwide for 36 cancers in 185 countries. CA: Cancer J. Clin. **68**(6), 394–424 (2018)
2. Lee, H., Kim, J.: Segmentation of overlapping cervical cells in microscopic images with superpixel partitioning and cell-wise contour refinement. In: Computer Vision and Pattern Recognition (CVPR) workshop on Computer Vision for Microscopy Image Analysis (CVMI) (2016)
3. Zack, G.W., Rogers, W.E., Latt, S.A.: Automatic measurement of sister chromatid exchange frequency. J. Histochem. Cytochem. **25**(7), 741–753 (1977)
4. Zhang, L., Kong, H., Chin, C.T., Liu, S., Chen, Z., Wang, T., Chen, S.: Segmentation of cytoplasm and nuclei of abnormal cells in cervical cytology using global and local graph cuts. Comput. Med. Imaging Graph. **38**(5), 369–380 (2014)

5. Gençtav, A., Aksoy, S., Onder, S.: Unsupervised segmentation and classification of cervical cell images. Pattern Recognit. **45**(12), 4151–4168 (2012)
6. Li, K., Zhi, L., Liu, W., Yin, J.: Cytoplasm and nucleus segmentation in cervical smear images using radiating GVF Snake. Pattern Recognit. **45**(4), 1255–1264 (2012)
7. Bergmeir, C., García Silvente, M., Manuel Benítez, J.: Segmentation of cervical cell nuclei in high-resolution microscopic images: a new algorithm and a web-based software framework. Comput. Methods Programs Biomed. **107**(3), 497–512 (2012)
8. Nosrati, M.S., Hamarneh, G.: Segmentation of overlapping cervical cells: a variational method with star-shape prior. In: IEEE International Symposium on Biomedical Imaging (ISBI), pp. 186–189 (2015)
9. Lu, Z., Carneiro, G., Bradley, A.P.: Automated nucleus and cytoplasm segmentation of overlapping cervical cells. In: Springer International Conference on Medical Image Computing and Computer-Assisted Intervention, pp. 452–460 (2013)
10. Song, Y., Zhang, L., Chen, S., Ni, D., Lei, B., Wang, T.: Accurate segmentation of cervical cytoplasm and nuclei based on multiscale convolutional network and graph partitioning. IEEE Trans. Biomed. Eng. **62**(10), 2421–2433 (2015)
11. Vese, Luminita A., Chan, Tony F.: A multiphase level set framework for image segmentation using the Mumford and Shah model. Int. J. Comput. Vis. **50**(3), 271–293 (2002)
12. Donoser, M., Bischof, H., Wiltsche, M.: Color blob segmentation by MSER analysis. In: IEEE International Conference on Image Processing, pp. 757–760 (2006)
13. Overlapping cervical cytology image segmentation challenge—ISBI 2014. http://cs.adelaide.edu.au/~carneiro/isbi14_challenge, Accessed 24 Sept 2017

Macroscopic Reconstruction for Histopathology Images: A Survey

Bijoyeta Roy and Mousumi Gupta

Abstract Over the past decade, due to a dramatic increase in diseases, histopathology has become a vital part of medical science. Histopathology involves examining of tissue under a microscope to examine the manifestation of diseases. The term histology deals with the study of entire cells of the sample specimen, whereas histopathology is a technique, which deals with analyzing for microscopic changes or abnormalities in tissues that are caused as a result of diseases. With the computational advancement, tissue histopathology slides can be digitized with the help of whole slide digital scanners. Histopathology is a powerful diagnosis method to analyze the tissue structure. However, the microscopic images generally produce two-dimensional views. With the advent of new computer algorithms, 3D reconstruction from 2D histopathology images is a routine technique. However, the already developed algorithms exhibit distorted image and recur information loss, which is unacceptable for the pathologists. An accurate high-resolution 3D reconstruction for histopathology sections will lead to a better diagnosis. This paper describes the 3D visualization techniques and its applicability in histopathology.

Keywords Histopathology · Medical imaging · 3D reconstruction · Confocal microscopy · Tissue histopathology · Histology sections

1 Introduction

Over the past decade, due to dramatic advances in computational power, it has become possible to identify and classify diseases with computer-aided image analysis [1, 2]. Remarkable improvements in image analysis algorithms have allowed the development of powerful computer-assisted analytical approaches to radiological data

B. Roy (✉) · M. Gupta
Sikkim Manipal Institute of Technology, Sikkim Manipal University, Majitar, India
e-mail: bijoyeta.r@smit.smu.edu.in

M. Gupta
e-mail: mousumi.g@smit.smu.edu.in

M. Gupta et al. (eds.), *Computer Vision and Machine Intelligence in Medical Image Analysis*, Advances in Intelligent Systems and Computing 992,
https://doi.org/10.1007/978-981-13-8798-2_11

[3–5]. In the past few years, macroscopic view through 3D image reconstruction has gained enormous importance in various fields of digital image processing techniques, especially in biomedical imaging [6–8]. In the modern digitized world, the sample specimen for diagnosis is converted to 2D digitized form, and then they are studied thoroughly by the pathologists to identify the disease [9, 10]. Structural and functional elements for the tissue images can be highlighted through 3D reconstruction from 2D histopathology images [11], and this view provides greater accuracy in disease diagnosis. 3D image reconstruction is implemented in various tomographical modalities, which basically does imaging by sections or sectioning through the use of penetrating wave and has shown fruitful results [12–14]. In the previous study by various researchers, it is found that 3D reconstruction of X-ray images and magnetic resonance images (MR) images had shown a good result and contributed potentially in clinical decision-making. Geometric reconstruction of 3D anatomy from 2D X-ray images had helped the clinicians to identify and diagnose the disease in a more precise way. However, still, there exist various loopholes for histopathology images 3D reconstruction. Through efficient algorithm development, this 3D reconstruction for tissue structure can be improvised and a cost-effective tool can be developed for pathologists. Recent medical science is completely based on the result of biopsy of tissue samples. Histological sectioning serves as one of the most powerful methods for accurate high-resolution representation of tissue structures. The histological sections are viewed through a microscope and its corresponding two-dimensional images are created. However, it is very important to have ideal, nondestructive cross-sectional imaging techniques. A 2D histopathology image can be reconstructed into 3D and thus will reveal hidden microstructures in human histopathology tissues. The main goal of this paper is to highlight the importance of 3D reconstruction for histopathology images and discuss the various techniques that are applied so far in this field and their recent challenges.

This paper is structured to follow the general image analysis procedure for histopathology imagery. The paper mainly focused on 3D reconstruction of microscopic images. Section 2, discussed about the related literature regarding the facts and findings of various researchers on three-dimensional reconstruction and also depicted the basic workflow of 3D reconstruction technique. In Sect. 3, the importance of 3D reconstruction which facilitates better diagnosis of tissue samples is discussed. Sections 4 and 5 discuss about the various reconstruction techniques and tools which has proved to produce significant results. Finally, this paper concluded in Sect. 6.

2 Previous Related Works

Arakeri and Reddy in 2013 [15] proposed an approach to 3D reconstruction of brain tumor and estimation of its volume from a set of two-dimensional (2D) cross-sectional magnetic resonance (MR) images of the brain. They did their work by developing methods for segmentation, inter-slice interpolation, mesh generation, and simplification. Ignao and Carreras [16] proposed a method for simultaneous morphological and molecular analysis of thick tissue cells of the mammary gland and was successful in overcoming the penetration limitations of other microscopy modalities resulting in better 3D views of mammary gland. The 3D reconstruction includes rigid and nonrigid image registration, automatic segmentation, contour grouping, and volume rendering. Qeethara Kadhim Al-Shayea and Muzhir Shaban Al-Ani in 2016 [17] developed an approach of 3D reconstruction of an X-ray image using the concept of volume reconstruction by finding the contour of the given object in each slice, and then merging the contours to reconstruct the 3D objects. Annedore Punge et al. proposed an approach was discussed, which allowed a 3D image nanoscopy with a resolution, less than 80 nm using stimulated emission depletion (STED) microscopes and photo-switching microscopy methods. STED microscopy made it possible to do imaging with focused visible light at a resolution of 20–50 nm and it showed better results as compared to confocal microscopy [18]. From various studies, it was revealed that 3D optical microscopic techniques like deconvolution, confocal microscopy, etc., places severe constraints on the maximum thickness of a specimen that can be imaged [19, 20]. Sharp et al. [21] developed a method where optical projection tomography was used to produce high-resolution 3D images of biological samples with thickness up to 15 nm. Though traditional confocal microscopy attempts to minimize the noise from out of focus regions by illuminating only those points on the focal plane, however, it is limited to imaging only fluorescent signals. In 2008, K. Becker, Jahrling et al. [22] presented a concept, which produces a 3D view of medical specimens that can be reconstructed using a microscopical technique called ultramicroscopy which allows optical sectioning of samples. In this paper [22], the principle of light sheet illumination was used to get an ultramicroscopic 3D view of whole immunohistochemically labeled mouse embryos and adult Drosophila giving in-depth sight to their anatomy. This technique proved to be better than confocal microscopy, which can deal with specimen size limited to few 100 micrometers and therefore can give limited fields of view and penetration depths of less than 1 mm. Ultramicroscopy made it possible to make 3D reconstructions of macroscopic specimen in the millimeter or centimeter range with micrometer resolution [22, 23]. In 2009, the authors have pointed out that histopathology slides provide a more comprehensive view of diseases as compared to cytology imagery because the preparation process preserves the underlying tissue architecture. Histopathology deals with the study of tissue structure through a microscope for surgical specimen, whereas cytology deals with the cell structure and their chemistry [24].

For the pathologists, it is necessary to visualize different histological sections to draw a conclusion about a particular disease. Macroscopic view or 3D view for histology specimen with different stains allows visualizing the spatial and structural alignment for tissue sections. Formalin-fixed and paraffin-embedded tissue blocks and their multiple sections are stained with H & E (Hematoxylin and Eosin). H & E are standard stains, which are used to highlight the nuclei and cytoplasm to visualize the structure in the tissue [24]. The fundamental steps for 3D reconstruction from 2D histopathology are shown in Fig. 1.

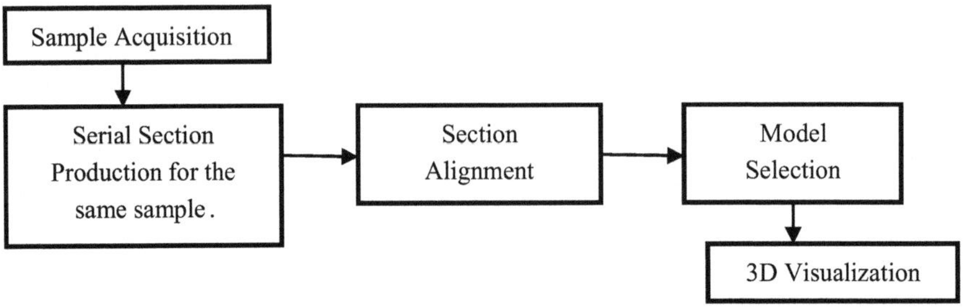

Fig. 1 Fundamental steps for 3D reconstruction of histopathology images

The acquisition for histopathology sections is obtained individually for each section. Quality of serial sections produced has a significant role in successful 3D visualization of biological structures. 3D reconstruction involves registering each section to get the original alignment. The next step is to align the serial sections properly so that it helps to retrieve maximum information about the internal structures of the sample specimen. Then based on the type of sample, we need to select a particular model that fits best for that biological specimen for producing 3D visualization.

3 Importance of 3D Reconstruction of Microscopic Images

Histopathology deals with examining the biological tissue samples to detect the presence of diseased cells or abnormalities in tissues in very fine details. It deals with the visualization of tissue and microscopic features of cells by examining tissues under light microscopes. However, the 2D images that are captured by the microscopes cannot give every minute details of the sample. Thus, generating a macroscopic or three-dimensional views for tissue structure of the pathology specimen will produce the actual mechanism of disease. 3D view for tissue can provide a comprehensive visualization of molecular anatomy. The main advantage of 3D reconstruction for histopathology images is the opportunity to study each spectrum of disease progression with an appropriate quantification.

Figure 2 depicts the whole procedure of macroscopic visualization of a histopathology section. The input is a histopathology image of colon adenocarcinoma. The image includes three parts such as stroma, mucous, and tumor. Tumorous region is an object of interest. A serial section for this tumorous region has to retrieve for macroscopic visualization. An image registration algorithm is required for exact alignment of these serial sections. Further, a reconstruction technique will provide the macroscopic view for the extracted portion. This view will provide the exact mechanism of cancer giving an insight into morphological as well as molecular analysis of tissue samples.

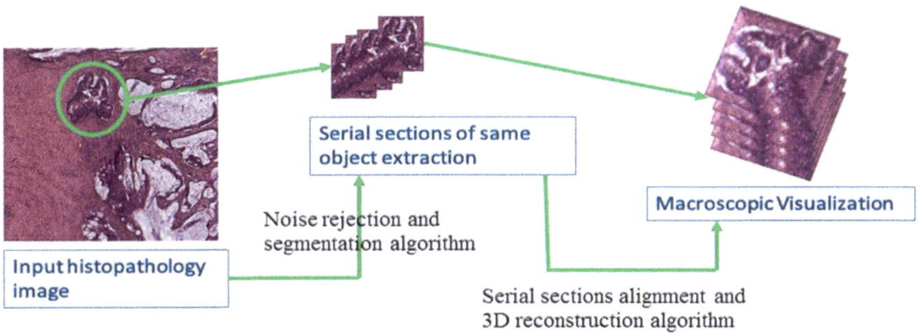

Fig. 2 Histopathology image for colon adenocarcinoma [taken from Pathology department of Sikkim Manipal Institute of Medical Sciences]

4 Fundamental Applied Algorithms for 3D Reconstruction

3D reconstruction of histopathology images is generally based on two basic algorithmic techniques. These two techniques worked well in generating a good reconstruction of the sample specimen and provided appealing 3D visualization as compared to 2D visualization.

4.1 3D Visualization By Volume Reconstruction

In 2016, the authors have proposed a technique of 3D reconstruction using the concept of volume reconstruction. Volume reconstruction is an important concept in the field of image visualization [17].

In this paper, the method that was proposed deals with 3D visualization from 2D images that included various objects. This proposed algorithm of image visualization is used to generate 3D objects depending on the extraction of the important features. The method followed in the proposed method is as follows:

(i) First, various slices of the same section are obtained;
(ii) The contour of the interested areas are found in each image;
(iii) The desired zone in each image is specified by a rectangle;
(iv) The distance between the two images is calculated (D);
(v) The volume for the two images is calculated by the given formula

$$V_n = \frac{1}{3} D(S1 + \sqrt{S1S2} + S2) \tag{1}$$

(vi) The volume is reconstructed by adding the distances evaluated between the thin slices of images (Fig. 3).

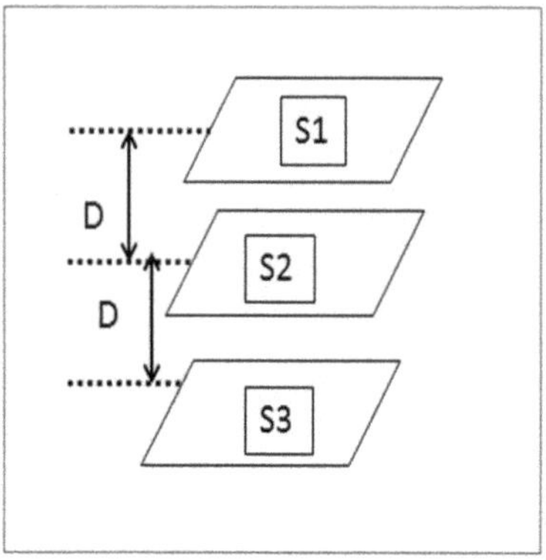

Fig. 3 Distance calculation for reconstruction of volume from slices

Al-Shayea and Al-Ani et al. [17] applied this technique for medical image 3D reconstruction and Arganda-Carreras et al. [16] applied this methodology for histology image reconstruction and produced a significant result.

4.2 Three-Dimensional Reconstructions Using Multistained Registration

Song et al. [25] applied a 3D reconstruction technique on two surgical liver specimen images of differently stained sections. The main purpose of staining is to reveal the cellular components of surgical specimens. This technique used the concept of multistained registration for the alignment of the images. Registration is done on the thin slices stacking sequentially one after another and then aligning it. After this registration, the set of aligned images can be concatenated to display the three-dimensional structure. However, the problem of different dissimilar structure appearances of adjacent tissue sections might arise in the case of multistained registration. This issue can be addressed using the technique of unsupervised content classification, which has the capability of identifying common content class from multistained image pair. This technique is capable of identifying the problematic sections and performs well on serial sections having a continuous cut, which is an important step for histopathology image reconstruction. The main steps involved in multistained registration are as follows:

(i) Rigid registration which results in roughly aligned pair of images;
(ii) Content classification;
(iii) Nonrigid registration based on probability image pairs resulting in improved aligned color image pair;
(iv) Resolution is changed (increased) and then steps (ii) and (iii) are repeated to get a better result (Fig. 4).

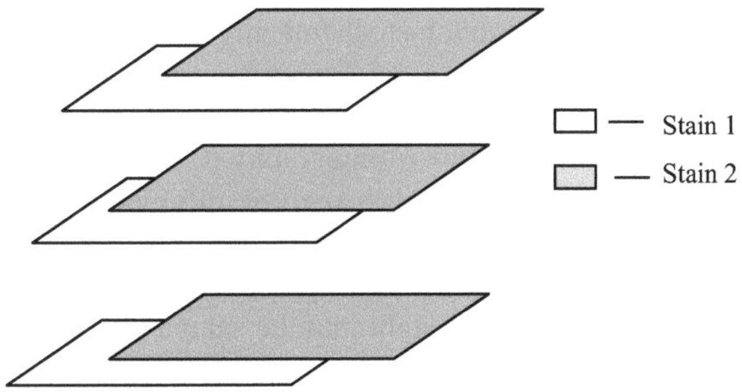

Fig. 4 Multistained registration proposed by Song et al. [25]

5 Available Tools for 3D Reconstruction of Histopathology Images

5.1 Confocal Microscopy

Confocal microscopy has gained acceptance as an important technology of imaging cell structure and its internal topology. It is an important tool for quantifying and visualizing a three-dimensional structure. This method optically sections the specimen, which is noninvasive and the samples can be studied with better clarity. Since the optical sectioning can be done vertically (xz and yz planes) as well as laterally (x- and y-axis) and can be digitized, so it helps in getting a clear picture in terms of contrast, clarity, and detection sensitivity [11]. However, in previous researches, the authors have highlighted some drawbacks of confocal microscopy as compared to their proposed method. Though this method provides high-resolution imaging of cell distribution, however, from [11, 16], we can summarize that though confocal microscopy proved to be a powerful tool for giving a 3D view of microscopic image, however, it has limitations with respect to penetration depth and sample specimen coverage area.

5.2 Ultramicroscopy

Ultramicroscopy is an advanced investigative medical tool that allows to generate a three-dimensional view by optical sectioning of biological specimens [22]. In various literatures, the authors have discussed that ultramicroscopy is a powerful and effective tool for 3D image reconstruction. It has proved to be a novel technique, which has given new impulses in the field of biological science. It can reconstruct 3D architecture of very minute specimen of size in cm with a resolution in micrometers. And hence, it has gained its importance and considered as a versatile tool for studying the anatomy of numerous biological and medical specimens. This technique is used in wide areas like viewing the neuron structure of the brain of a mouse, reconstruction of whole immunohistochemically labeled mouse embryos, giving 3D view to microvascular networks, etc. Basically, it can be thought that ultramicroscopy acts as a bridge for the gap between traditional confocal microscopy and other macroscopic techniques like computer tomography (CT), etc. From [22], it can be said that ultramicroscopy is not limited to analyze millimeter-sized histopathological samples, it can cover centimeter-sized specimen sample with a penetration depth of more than 1 mm. It is ideal for the investigation of disease models as the 3D view that it creates gives a detailed insight into the sample and allows high-throughput phenotype screening [23].

5.3 Surface Imaging Microscopy

Surface image microscopy (SIM) is an advanced tool that allows generating volumetric images of high resolution of biological samples. It can create three-dimensional views of sample of the embryo at high contrast and high resolution. It is capable of providing more precise three-dimensional views of the embryo of a chick. Also, it has proved to give excellent results of the imaging while dealing with three different vertebrate systems like mouse chicken and frog. There are various techniques, which provide tools for producing high-resolution image of small volumes or low-resolution images of high volume. However, SIM can be used to bridge this gap [26, 27]. The main steps involved in this technique are the following:

(a) It first labels, infiltrates, and then embeds a fluorescently labeled specimen in a black polymer and then loads the sample into a motorized translation stage.
(b) Then, a diamond knife is used to remove a thin section of the sample and places it over objective lens for view.
(c) Finally, the image of the sample is captured by a CCD camera and the same process is repeated again. These aligned series of 2D raw images can be computationally combined into a 3D structure.

This technique can be used over very delicate histopathological samples as well as over rigid samples like bones. Ewald et al. [26] proposed SIM technique in three different categories of vertebrates and concluded that it can produce 3D images with high resolution and high contrast. Also, its performance was better with respect to confocal microscopy in terms of penetration depth and coverage of sample size.

5.4 Optical Projection Tomography

Optical Projection Tomography is a microscopic technique, which is used to create three-dimensional views of small biological tissues. With respect to resolution, it gives better results as compared to magnetic resonance imaging (MRI). It can operate on both fluorescent and nonfluorescent biological samples with a penetration depth (thickness) of 15 mm [21].

5.5 *Most Commonly Used Microscopy*

To visualize histologic and histomorphometric changes, the routine diagnostic tool used by the pathologists is light microscopy. Pathologists recommend treatment after viewing the abnormal matrix characteristics from different thin sections of the sample specimen. The diagnosis is mostly based on reviewing the tissue pattern. Light microscopy provides high-resolution images. Pathologists were able to distinguish different tissue structure through these high-resolution images. A macroscopic view for tissue images will provide an extra benefit for disease diagnosis in the field of histopathology.

A comparison analysis for the mentioned 3D reconstruction tools of histopathology images provides a brief overview as shown in Table 1.

Table 1 Comparison of various 3D visualization tools

3D visualization tools	Advantages	Loopholes	Penetration depth and coverage area	Area of application	References
Confocal microscopy	Excellent spatial resolution Enhanced Signal-to-noise ratio	Restricted to samples stained with fluorescent probes and it is very costly (10 times), slow scan process	Penetration depth less than 1 mm and works on small sample specimen (few 100 micrometers)	Surface of embryo, Meibomian gland disfunction, DNA hybridization, etc.	[4, 11, 16]
Ultramicroscopy	Provides micrometer resolution (1–2 μm), less acquisition time (20–40 min)	Shrinkage of Tissue	Applied to large specimen sample having size in cm and allows deeper penetration	Imaging of whole mouse embryos and flies, imaging neuronal networks in the whole brain, malignant tissue, vascular networks, etc.	[22, 23]

(continued)

Table 1 (continued)

3D visualization tools	Advantages	Loopholes	Penetration depth and coverage area	Area of application	References
Surface imaging microscopy	High sensitivity and contrast, supports samples of varying density, isotropicity	Cannot image antibody revealed protein expression, green fluorescent protein expression in intact embryos, etc.	Provides cellular resolution on large samples	Embryo samples (Chick, mouse, and frog embryos, etc.), resin embedded tissues, etc.	[26, 27]
Optical projection tomography	Maximize depth of focus, supports molecular and morphological imaging	Does not provide sufficient colored stains, Limited to a resolution of about 20 μm, processing and imaging is time consuming	Supports sample thickness of around 5–10 mm	Imaging embryonic organs, analysis of altered morphology in mutant mice, lymphoid tissues, organogenesis, etc.	[21]

6 Conclusions

This paper presented the developed algorithmic techniques and tools used in 3D reconstruction of histopathology images. As per the author's knowledge, obtaining 3D view for histopathology images is still challenging. The already developed reconstruction techniques do not give high-resolution result for all types of sample specimens. This paper also represented the advantages of 3D reconstruction over 2D histopathology images. After studying various previous literatures, a conclusion can be made that the major gap in algorithm development for 3D reconstruction is an alignment of histology sections for the same sample. As well as the selection of correct 3D reconstruction technique is also a major concern for getting high-resolution and high-contrast macroscopic or 3D view.

Acknowledgements The Department of Pathology at Sikkim Manipal Institute of Medical Sciences provided support for histopathology image.

References

1. Tang, J., Rangayyan, R., Xu, J., Naqa, I. E., Yang, Y.: Computer aided detection and diagnosis of breast cancer with mammography: recent advances. IEEE Trans. Inf. Technol. Biomed. **13**(2), 236–251(2009)
2. Dauguet, J., Delzescaux, T., Condé, F., Mangin, J.F., Ayache, N., Hantraye, P., Frouin, V.: Three-dimensional reconstruction of stained histological slices and 3D non-linear registration with in-vivo MRI for whole baboon brain. J. Neurosci. Methods **164**(1), 191–204 (2007)
3. Lopes, S., Jayaswal, D.: A methodical approach for detection and 3-D reconstruction of brain tumor in MRI. Int. J. Comput. Appl. **118**(17), 37–43 (2015)
4. Capek, M., Bruza, P., Janacek, J., Karen, P., Kubınova, L., Vagnerova, R.: Volume reconstruction of large tissue specimens from serial physical sections using confocal microscopy and correction of cutting deformations by elastic registration. Microsc. Res. Tech. vol 72, pp. 110–119 (2009)
5. Markelj, P., Tomazevic, D., Likar, B., Pernus, F.: A review of 3D/2D registration methods for image-guided interventions. Med. Image Anal. **16**, 642–661 (2012)
6. Ehlke, M., Ramm, H., Lamecker, H., Hege, H., Zachow, S.: Fast generation of virtual X-ray images for reconstruction of 3D anatomy. Trans. Visual. Comput. Graph. **19**(12), 2673–2782 (2013)
7. Kagawa, K., Lee, W.R., Schultheiss, T.E., Hunt, M.A., Shaer, A.H., Hanks, G.E.: Initial clinical assessment of CT-MRI image fusion software in localization of the prostate for 3D conformal radiation therapy. Int. J. Radiat. Oncol. **38**, 319–325 (1997)
8. Dey, D.: Mixing reality merging endoscopic images and 3D surfaces. In: Third International Conference on Medical Image Computing and Computer-Assisted Intervention (MICCAI), pp. 796–805, Pittsburgh, Pennsylvania, USA, LNCS 1935 (2000)
9. Fernandez, D.C., Bhargava, R., Hewitt, S.M., Levin, I.W.: Infrared spectroscopic imaging for histopathological recognition. Nature Biotechnol. **23**, 469–74 (2005)
10. Hosseinian, S., Arefi, H.: 3D reconstruction from multi-view medical X-ray images—Review and evaluation of existing methods. In: International Conference on Sensors & Models in Remote Sensing & Photogrammetry, pp. 23–25 (2015)
11. Ourselin, S., Roche, A., Subsol, G., Pennec, X., Ayache, N.: Reconstructing a 3D structure from serial histological sections. In: Image and vision computing, International Workshop on Biomedical Image Registration, pp. 25–31, Bled, Slovenia, vol. 19 (2001)
12. Brune, R.M., Bard, J. B. L, Dubreuil, C., Guest, E., Hill, W., Kaufman, M., Stark, M., Davidson, D., Baldock R.A.: A three-dimensional model of the mouse at embryonic day 9, Dev. Biol. **216**, 457–468 (1999)
13. Lixu, G., Peters, T.: 3D segmentation of medical images using a fast multistage hybrid algorithm. Int. J. CARS **1**, 23–31 (2006)
14. Zhang, Q., et al.: Rapid voxel classification methodology for interactive 3D medical images visualization. In: MICCAI 2007, the 10th International Conference on Medical Image Computing and Computer Assisted Intervention, pp. 86–93 (2007)
15. Arakeri, M.P., Reddy, G.R.M., Mohana, R.: An effective and efficient approach to 3D reconstruction and quantification of brain tumor on magnetic resonance images. Int. J. Signal Process. Image Process. Pattern **6**, 111–128 (2013)
16. Carreras, I.A., Ferna, R., Lez, N., Mun, A., Barrutia, O., Solorzano, C.O.: 3D reconstruction of histological sections: application to mammary gland tissue. Microsc. Res. Tech. **73**, 1019–1029 (2010)
17. Al-Shayea, Q.K., Al-Ani, M.S.: An efficient approach to 3d image reconstruction. Int. J. Comput. Sci. Netw. Secur. **16**, 35–42 (2016)
18. Punge, A., Rizzoli, S.O., Jahn, R., Wildanger, J.D.: 3D reconstruction of high-resolution STED microscope image. Microsc. Res. Tech. **71**, 644–650 (2008)
19. Al-Shayeh, K., Al-Ani, M.S.: Efficient 3D object visualization via 2D images. Int. J. Comput. Sci. Netw. Secur. **9**, 234–239 (2009)

20. Sugawara, Y., Kamioka, H., Honjo, T., Yamamoto, T.: Three-dimensional reconstruction of chick calvarial osteocytes and their cell processes using confocal microscopy. Bone **36**, 877–883 (2005)
21. Sharpe, J., et al.: Optical projection tomography as a tool for 3D microscopy and gene expression studies. Science **296** (2002)
22. Becker, K., Jahrling, N., Kramer, E.R., Schnorrer, F., Dodt, H.U.: Ultramicroscopy: 3D reconstruction of large microscopical specimens. Biophoton **1**, 36–42 (2008)
23. Dodt, H.U., Leischner, U., Schierloh, A., Jahrling, N., Mauch, C.P., Deininger, K., et al.: Ultramicroscopy: three-dimensional visualization of neural networks in the whole mouse brain. Nat Methods **4**, 331–336 (2007)
24. Gurcan, M.N., Boucheron, L., Can, A., Madabhushi, A., Rajpoot, N., Yener, B.: Histopathological image analysis: a review. IEEE Rev. Biomed. Eng. **2** (2009)
25. Song, Y., Treanor, D., Bulpitt, A.J., Magee, D. R.: 3D reconstruction of multiple stained histology images. J. Pathol. Inform. **4** (2013)
26. Ewald, A.J., Mcbride, H., Reddington, M., Fraser, S.E., Kerschmann, R.: Surface imaging microscopy-an automated method for visualizing whole embryo samples in three dimensions at high resolution. Dev. Dyn. **225**, 369–375 (2002)
27. Gerneke, D.A., Sands, G.B., Ganesalingam, R., Joshi, P., Caldwell, B.J., Smaill, B.H., Legrice, I.J.: Surface imaging microscopy using an ultramiller for large volume 3D reconstruction of wax-and resin-embedded tissues. Microsc. Res. Tech. **70**, 886–894 (2007)

Likelihood Prediction of Diabetes at Early Stage Using Data Mining Techniques

M. M. Faniqul Islam, Rahatara Ferdousi, Sadikur Rahman and Humayra Yasmin Bushra

Abstract Diabetes is one of the fastest growing chronic life threatening diseases that have already affected 422 million people worldwide according to the report of World Health Organization (WHO), in 2018. Due to the presence of a relatively long asymptomatic phase, early detection of diabetes is always desired for a clinically meaningful outcome. Around 50% of all people suffering from diabetes are undiagnosed because of its long-term asymptomatic phase. The early diagnosis of diabetes is only possible by proper assessment of both common and less common sign symptoms, which could be found in different phases from disease initiation up to diagnosis. Data mining classification techniques have been well accepted by researchers for risk prediction model of the disease. To predict the likelihood of having diabetes requires a dataset, which contains the data of newly diabetic or would be diabetic patient. In this work, we have used such a dataset of 520 instances, which has been collected using direct questionnaires from the patients of Sylhet Diabetes Hospital in Sylhet, Bangladesh. We have analyzed the dataset with Naive Bayes Algorithm, Logistic Regression Algorithm, and Random Forest Algorithm and after applying tenfold Cross- Validation and Percentage Split evaluation techniques, Random forest has been found having best accuracy on this dataset. Finally, a commonly accessible, user-friendly tool for the end user to check the risk of having diabetes

M. M. F. Islam (✉)
Queen Mary University of London, 153A Ilford Lane, Ilford, London IG12RR, United Kingdom
e-mail: m.islam@smd17.qmul.ac.uk

R. Ferdousi (✉)
Metropolitan University Sylhet, 131/8, Tootpara Taltola Hospital Road Khulna, Sylhet, Bangladesh
e-mail: rahatara@metrouni.edu.bd

S. Rahman
Metropolitan University Sylhet, House no-11, Road no-02, East Saplabag Tilagor, Sylhet, Bangladesh
e-mail: rahmansadik004@gmail.com

H. Y. Bushra
Metropolitan University Sylhet, 36/A Sharnali, Syed Manjil, Vartokhala, Sylhet, Bangladesh
e-mail: humayrabushra234@gmail.com

M. Gupta et al. (eds.), *Computer Vision and Machine Intelligence in Medical Image Analysis*, Advances in Intelligent Systems and Computing 992,
https://doi.org/10.1007/978-981-13-8798-2_12

from assessing the symptoms and useful tips to control over the risk factors has been proposed.

Keywords Diabetes risk · Symptom · Early stage · Data mining · KDD · Dataset · Evaluation model · Supervised learning algorithms · Unsupervised learning algorithms · Dataset · Mining tools

1 Introduction

Diabetes Mellitus, a chronic metabolic disorder, is one of the fastest growing health crises of this era regardless of geographic, racial, or ethnic context. Commonly, we know about two types of diabetes called type 1 and type 2 diabetes. Type 1 diabetes occurs when the immune system mistakenly attacks the pancreatic beta cells and very little insulin is released to the body or sometimes even no insulin is released to the body. On the other hand, type 2 diabetes occurs when our body doesn't produce proper insulin or the body becomes insulin resistant. Some researchers divided diabetes into Type 1, Type 2, and gestational diabetes [1]. Gestational diabetes is a type of diabetes which occurs only in pregnancy due to hormonal changes. The common symptoms of diabetes are polyuria, polydipsia, polyphagia, sudden weight loss (usually Type 1), weakness, obesity (usually Type 2), delayed healing, visual blurring, itching, irritability, genital thrush, partial paresis, muscle stiffness, alopecia, etc. [1, 2].

This could be a clear evidence that, according to WHO, the number of the diabetic patient had been sharply increased from 108 million in 1980 to 422 million in 2014 [3]. The most alarming fact is that more than 80% of diabetic people were from low- and middle-income countries in 2013 and the prevalence is surging up in these countries. Recently, Diabetes Australia has published that, Diabetes even may exist up to 7 years before clinical diagnosis [4], which was even up to 12 years previously noted by Harris et al. [5]. Within this time frame, people may gradually suffer from fatal complications like heart attacks, strokes, eye damage resulting in blindness, foot ulcer, amputation of the affected limb, kidney damage, and other forms of multi-organ damage [5]. Most of the cases, these complications would be easily controlled or even prevented in some cases with early detection and treatment initiation that could possibly save around 1415 AUD [4]. The degree of diabetic complication is more when the period between onset of disease and treatment initiation is longer [5]. According to Diabetes Australia, failure in early detection of TYPE 2 could cost the Australian healthcare system more than 700 million dollars each year [4]. In 2017, the total expenditure of diagnosed diabetes in the United States alone was 327 billion USD [2]. In [6], in the year 2011, China had experienced 90 million (9% of the population), India had 61.3 million (8% of the population) and Bangladesh had 8.4 million (10% of the population). Comparing to developed countries like Australia and The USA, low and middle-income countries cannot afford the burden of managing such a costly disease like diabetes, the prevalence of which is increasing at an alarming rate. Therefore, early diagnosis and initiation of appropriate therapeutic management may play a pivotal role in the patient outcome and reduce the gross national expenditure and production loss. Another considering issue is that globally, OGTT (Oral Glucose Tolerance Test), HbA1c are widely accepted methods of diagnosis of diabetes which are usually referred by the physician after developing patient's sign symptoms. However, these tests are not so cheap, lab reagent and technician dependent as well as time consuming, these tests are not available in remote settings. As the protocol of the treatment is not only long term but also expensive, the earlier detection of diabetes is beneficial in terms of patient's health, individual and national expenditure, as well as productivity [7].

In this modern era of technology, computer technology can help us to detect diseases accurately and can save our time and money. Data mining is an important field of computer science which is used for prediction. It is the process of discovering new data from previously known data through data analysis [5]. To predict a disease using data mining approaches, we need its symptoms along with clinical data. Symptoms are a very important factor for new patients and early stage prediction since they have no data except symptoms. We also need clinical data for analyzing and discovering new data.

Early assessment of symptoms can be possible by creating mass awareness, manual assessment by health workers/assistants (where doctor/facilities are not available due to remoteness) in the rural setting, or by some user-friendly and cost-effective system. This system should be designed for specific target users so that it is easily accessible for mass people. As a result, early diagnosis of diabetes, prediabetes, risk of diabetes through symptom assessment by any means not only can prevent fatal outcomes of diabetes but also can save up such a huge financial expenditure as stated above, as well as increase the national productivity level, which could bring fruitful outcomes in low- and middle-income countries. Thus in this paper, we are providing analysis on a newly created dataset of 520 instances using different data classification algorithm to find one that provides better accuracy. Finally, we have proposed a tool for the end users to predict the likelihood of diabetes risk at its early stage, using patients symptoms with the help of data mining techniques.

2 Literature Review

In this section, different research works that were envisioned to predict diabetes using data mining have been provided with their remarkable contribution.

In [8], the authors collected 865 data with 9 attributes called Sex, Diastolic B.P, Plasma glucose, Skin fold thick, BMI, Diabetes Pedigree type, No. of times Pregnant, 2 h Serum Insulin and Diabetes probability and used WEKA 3.6.6 for the experiment. They found 100% accuracy with J48 (C4.5), 98.48% with the Decision Tree, 97.85% with the Neural Network, 96.54% with JRip and 95.85% with Naive Bayes algorithm. They also calculated the performance over time.

In [9], the author used 738 patient's data for experimental analysis. To predict diabetes, they introduced algorithms like CNN, KNN, SVM, SVM+LDA, NB, SVM,

ID3, C4.5, CART for comparing the analysis on the dataset. The best accuracy at 88.10% was achieved using SVM and LDA algorithm together.

In [10], the author compared three machine learning algorithms to predict diabetes. They introduced SVM, Logistic regression, ANN to seven attributes of their data including the Glucose, Blood Pressure, Skin, Thickness, Insulin, BMI, Diabetes Pedigree Function, and the age. After comparing their features, the researcher opinionated that the Support Vector Machine (SVM) found SVM as the best classification method.

In [11], the author used Artificial Neural Network for predicting diabetes They collected 250 diabetes patients data from Pusat Perubatan University Kebangsaan Malaysia, Kuala Lumpur and between 25 and 78 years old. They used MATLAB to train data. They had done Regression analysis using different algorithms, BFGS Quasi-Newton, Bayesian Regulation, and Levenberg–Marquardt. They found 88.8% accuracy with Bayesian Regulation algorithm.

In [12], the authors used Pima Indian Diabetes dataset and WEKA as their software tool for dataset testing. They tested their dataset with Naive Bayes (NB), Random Forest (RF), and function-based Multilayer Perceptron (MLP) algorithms and used different test methods called FCV, PS, and UTD. They also predicted with preprocessed and without preprocessed data and made a convenient table on their result. They found 100% accuracy with Random Forest algorithm with UTD method. However, the author stated that preprocessed data can give more accuracy in the Naive Bayes algorithm.

In [13], the author has created a new model for type 2 diabetes patients treatment. He collected 318 medical records with 9 nominal attributes including the patient's Gender, Age, Smoking, History of hypertension, Renal problem, Cardiac problem, and Eye problem. The duration of Diabetes Basic control was used as a class level attribute. He used the J48 algorithm and found an accuracy rate of 70.8% and ROC (Receiver operating characteristic) rate was 0.624.

In [14], the authors have predicted diabetes with supervised and unsupervised learning. They used the software tool WEKA to find a better prediction algorithm in machine learning. Finally, they concluded that ANN or Decision tree is the best way for diabetes prediction.

In [15], the author used Logistic Regression to predict diabetes. In their data, they used Age, Smoking, Parental Diabetes Mellitus, Hypertension & Waist Circumference, Sex, BMI, and HBA1C information as the attribute. The data analysis was conducted using the software tool IBM SPSS 20.0. In result, they found the likelihood 78.5565%, Cox & Snell R Square Nagelkerke Square 0.628, and Nagelkerke R Square 0.839.

In [5], the author aimed to forecast whether the patient has been affected by diabetes or not using the data mining tools and the MV dataset. This dataset contains 1024 complete instances of 26 parameters. MV dataset was collected from various districts people using questionnaires. They experimented Decision Trees to predict diabetes for local and systemic treatment.

3 Proposed System Architecture

The proposed system architecture is shown in Fig. 1. The dataset containing the information about the symptoms of the patients will be fed to the prediction algorithms like Naive Bayes, Decision Trees, Logistic Regression, and Random forest algorithm. Then the performance of the algorithms will be tested with appropriate evaluation model, in particular, tenfold cross-validation and percentage split techniques. Then the best algorithm chooses to build the system for the end users using the dataset as database. Taking the symptom from the user as input, the system will support the user for risk prediction.

4 Methodology

The dataset was analyzed using the following classification algorithms. The data analysis procedure can be formulated according to the algorithm 1.

```
Data: Diabetes Symptom Dataset
Result: Classification Technique with Best Accuracy
i=1;
bestAccuracy=0;
maxAccuracy=accuracy(1st algo);
while i is less than or equal to numberOfAlgorithm do
  | if accuracy(ith algo) greater than maxAccuracy) then
  |   | maxAccuracy=accuracy(ith algo);
  |   | i++;
  | end
  | bestAccuracy=maxAccuracy;
end
```

Algorithm 1: Algorithm for Dataset Analysis

4.1 Naive Bayes (NB)

Naive Bayes uses a probabilistic algorithm. The algorithm assumes the features and variables provided are independent to one another. It is carried out by using a probabilistic approach, which determines class probabilities and predicts most probable classes. The following equation from (1) to (3) represent the classification formula, where Pos and Neg represent a person with diabetes risk and without diabetes risk, which are the values of the class attribute for this dataset. X is the instances of the dataset as well as person.

$$P(Pos|X) = P(x_1|pos) * P(x_2|pos) * \cdots * (x_n|pos) * P(Pos) \qquad (1)$$

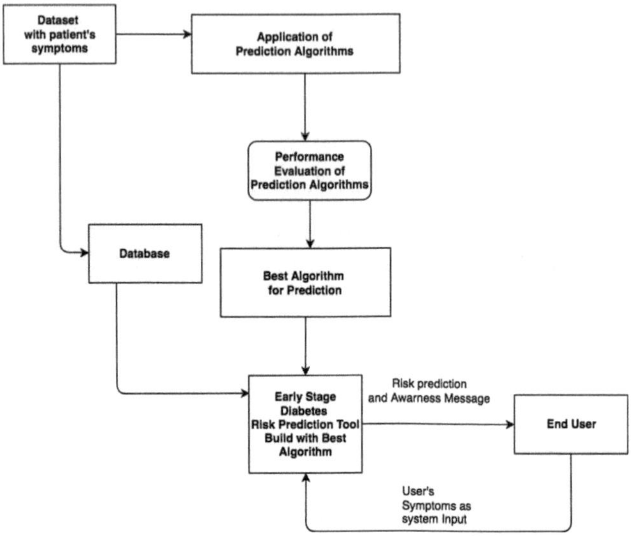

Fig. 1 Proposed system architecture

$$P(Neg|X) = P(x_1|neg) * P(x_2|neg) * \cdots * (x_n|neg) * P(Neg) \tag{2}$$

$$P(x_i|Pos) = \frac{(TotalPos|x_i)}{TotalPos} \tag{3}$$

where i is an increment until it reaches n(total attributes for our data).

4.2 J48 Decision Tree (J48 DT)

J48 algorithm is a kind of decision tree which belongs to the supervised learning algorithm. It is one of the most important classifiers as it is easy and simple to implement. Using the decision tree, a dataset is broken down into smaller and smaller subsets while at the same time an associated decision tree is incrementally developed. The algorithm uses Eqs. (4)–(6) to find information gain for our dataset to predict the outcome.

$$E(P) = -\sum_{j=1}^{n} \frac{|P_j|}{|P|} \log \frac{|P_j|}{|P|} \tag{4}$$

$$E(j|P) = \frac{|P_j|}{|P|} \log \frac{|P_j|}{|P|} \tag{5}$$

$$Gain(P, j) = E(P - E(j|P)) \tag{6}$$

P represents total instance, n represents total number of classes, and j represents total number of attributes in the dataset.

4.3 Logistic Regression (LR)

The LR classifier works with the class and uses multinomial logistic regression model with a ridge estimator. For k number of classes and for instances n with attributes m, the parameter matrix B can be calculated with the matrix given in Eq. (7).

$$B = m * (k - 1) \tag{7}$$

The probability for class j with the exception of the last class is stated in (8) and the last class probability given in (9).

$$P_j(X_i) = \frac{exp^{\sum_{j=1}^{k-1} X_i B_j}}{(1 + exp)^{\sum_{j=1}^{k-1} X_i B_j}} \tag{8}$$

$$P'_j(X_i) = \frac{1}{(1 + exp)^{\sum_{j=1}^{k-1} X_i B_j}} \tag{9}$$

Thus, the negative multinomial log-likelihood is

$$\begin{aligned} L = -\textstyle\sum_{i=1}^{n}[\Sigma_j^{k-1}(Y_{ij} * ln(P_j(X_i))) + (1 - \Sigma_{j=1}^{k-1})* \\ ln(1 - \Sigma_{j=1}^{k-1} P_j(X_i))] + ridge * B^2 \end{aligned} \tag{10}$$

In order to determine accuracy B, L is kept minimized as much as possible.

4.4 Random Forest (RF)

Random forest uses bagging method to train the dataset. For a training set of $X = x_1, \ldots, x_n$ and $Y = y_1, \ldots, y_n$, it selects random sample B times with replacement of the training set and fits trees to these samples. After training, it predicts unseen samples x' by averaging the predictions from all the individual regression trees on x' as shown in Eq. (11) and also by taking the majority vote in the case of classification trees.

$$\hat{f} = \frac{1}{B} \sum_{b=1}^{B} f_b(x') \tag{11}$$

5 Experimental Analysis

Dataset details and the result analysis are represented in this section.

5.1 Dataset Details

This dataset contains reports of diabetes-related symptoms of 520 persons. It includes data about peoples including symptoms that may cause diabetes. This dataset has been created from a direct questionnaire to people who have recently become diabetic, or who are still nondiabetic but having few or more symptoms. The data has been collected from the patients using direct questionnaire from Sylhet Diabetes Hospital of Sylhet, Bangladesh.

The data preprocessing has been conducted by handling the missing values following the technique of ignoring the tuples with incomplete values. After preprocessing,

Table 1 Description of dataset

	Number of attributes	Number of instances
Diabetes symptom dataset	16	520

Table 2 Description of attribute

Attributes	Values
Age	1.20–35, 2.36–45, 3.46–55,4.56–65, 6.above 65
Sex	1.Male, 2.Female
Polyuria	1.Yes, 2.No.
Polydipsia	1.Yes, 2.No.
Sudden weight loss	1.Yes, 2.No.
Weakness	1.Yes, 2.No.
Polyphagia	1.Yes, 2.No.
Genital thrush	1.Yes, 2.No.
Visual blurring	1.Yes, 2.No.
Itching	1.Yes, 2.No.
Irritability	1.Yes, 2.No.
Delayed healing	1.Yes, 2.No.
Partial paresis	1.Yes, 2.No.
Muscle stiffness	1.Yes, 2.No.
Alopecia	1.Yes, 2.No.
Obesity	1.Yes, 2.No.
Class	1.Positive, 2.Negative.

Table 3 Comparison of evaluation metrics using tenfold cross-validation and percentage split (80:20)

Evaluation metrics	Cross-validation				Percentage split			
	NB	LR	J48	RF	NB	LR	J48	RF
Total number of instances	500	500	500	500	100	100	100	100
Correctly classified instances	437	462	478	487	88	91	95	99
	87.4%	92.4%	95.6%	97.4%	88%	91%	95%	99%
Incorrectly classified instances	63	38	22	13	12	9	5	1
	12.6%	7.6%	4.4%	2.6%	12%	9%	5%	1%

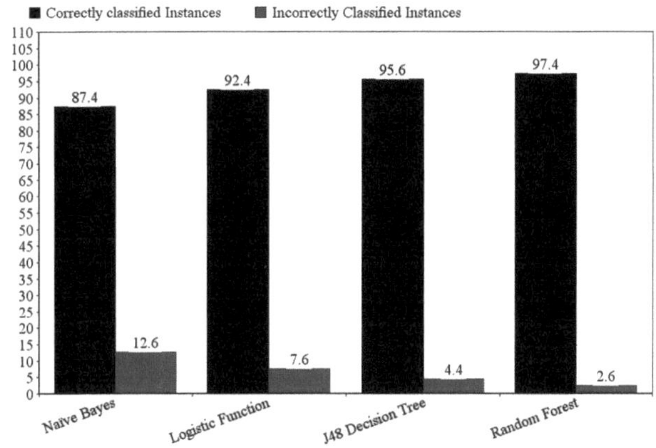

Fig. 2 Performance of classification algorithms using cross-validation technique

500 instances have been remained in total. Among them, 314 are positive values and 186 are negative values. The detail description of the dataset and the attributes are shown in Tables 1 and 2. Two class variables are used to find whether the patient is having a risk of diabetes (positive) or not (negative).

5.2 *Result Analysis*

Performance of different Data Mining techniques on our dataset with detailed accuracy information is represented in the following tables. Although Naive Bayes classifier is one of the most popular algorithms for data prediction, in case of our dataset, the accuracy of it was the lowest for both the cross-validation method and also for the percentage split. However, the best result was achieved using Random Forest Algorithm where using tenfold cross-validation 97.4% instances were classified correctly and using percentage split technique, it could classify 99% of the instances correctly as shown in Table 3. For the more semantic view of the performance of

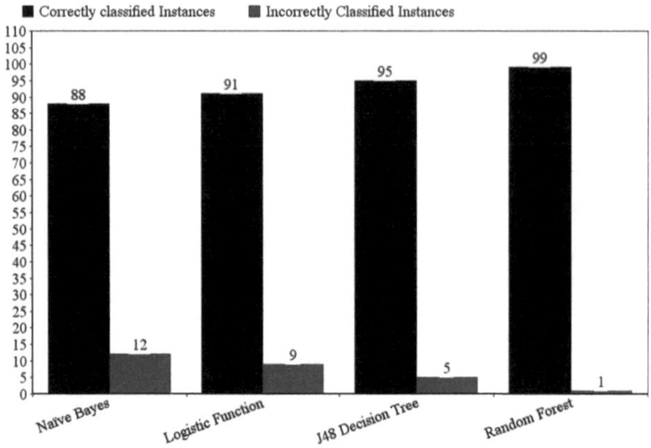

Fig. 3 Performance of classification algorithms using percentage split technique

Table 4 Comparison of performance parameters using tenfold cross-validation

Performance parameters	Class	Weighted average			
		NB	LR	J48	RF
TP rate	Positive	0.869	0.936	0.949	0.978
	Negative	0.886	0.903	0.968	0.968
	Weighted average	0.874	0.924	0.956	0.974
FP rate	Positive	0.118	0.097	0.032	0.032
	Negative	0.131	0.064	0.051	0.022
	Weighted average	0.123	0.084	0.039	0.029
Precision	Positive	0.925	0.942	0.980	0.981
	Negative	0.800	0.894	0.918	0.963
	Weighted average	0.879	0.924	0.957	0.974
Recall	Positive	0.869	0.936	0.949	0.978
	Negative	0.882	0.903	0.968	0.968
	Weighted average	0.874	0.924	0.956	0.974
F-measure	Positive	0.897	0.939	0.964	0.979
	Negative	0.839	0.898	0.942	0.965
	Weighted average	0.875	0.924	0.956	0.974

used algorithms using both evaluation techniques are depicted in graphs. In Fig. 2, the performance of the algorithms using cross-validation evaluation is depicted and in Fig. 3, the results from percentage split have been shown to represent the compar-

Table 5 Comparison of performance parameters using percentage split

Performance parameters	Class	Weighted average			
		NB	LR	J48	RF
TP rate	Positive	0.930	0.947	0.965	1.000
	Negative	0.814	0.860	0.930	0.977
	Weighted average	0.880	0.910	0.950	0.990
FP rate	Positive	0.186	0.140	0.070	0.023
	Negative	0.070	0.053	0.035	0.000
	Weighted average	0.136	0.102	0.055	0.013
Precision	Positive	0.869	0.900	0.948	0.983
	Negative	0.897	0.925	0.952	1.000
	Weighted average	0.881	0.911	0.950	0.990
Recall	Positive	0.930	0.947	0.965	1.000
	Negative	0.814	0.860	0.930	0.977
	Weighted average	0.880	0.910	0.950	0.990
F-measure	Positive	0.898	0.923	0.957	0.991
	Negative	0.854	0.892	0.941	0.988
	Weighted average	0.879	0.910	0.898	0.980

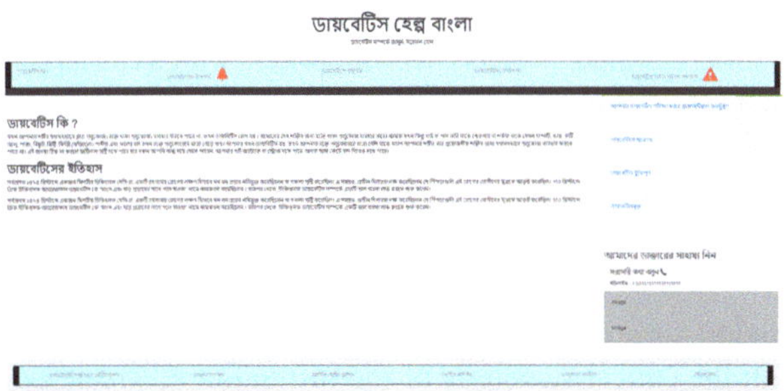

Fig. 4 Homepage of proposed tool

ative accuracy of the used algorithms. The comparison of detail performance using 10 fold cross validation and percentage split given in Tables 4 and 5 respectively.

6 Proposed Tool for the End Users

To provide instant help to the mass people for diabetes risk prediction, regardless of location age or educational background an easy and globally accessible system is required. As the web technology has quickly become the worlds most common way of searching data and services, a simple website could be undertaken to check the risk of the diabetes using users symptom as input. This website should provide both predictions of likelihood of having diabetes and some useful health tips for both the diabetic and nondiabetic. Useful health tips for a nondiabetic can reduce or delay the risk of him/her to have diabetes. A demo homepage of our proposed tool is shown in Fig. 4

7 Conclusion

The potentiality of diabetes is increasing among people of all age. The present study says that detection of diabetes at its early stage can play a pivotal role in treatment. Simple awareness measures such as low sugar diet, regular physical activity, and healthy lifestyle can avoid obesity. As the data mining methods, techniques and tools are becoming more promising to predict diabetes and eventually number of patients reduce the treatment cost, its role in this medical health care is undeniable. The main contribution is to find out the best algorithm for the prediction of newly created datasets made for diabetic risk prediction. We found that the Random Forest algorithm had performed with the best accuracy in percentage split evaluation test. Finally, a tool for the marginal user has been proposed, which can be used for diabetic risk prediction, awareness creation, and instant help. However, this research can be updated regularly with a dataset with more instances and can apply other widely accepted other data mining technologies for prediction purpose. As the system has been only prototyped, a deploying version of this system can be considered as a sustainable outcome of this research.

Ethical Approval: All procedures performed in studies involving human were in accordance with the ethical standards of the institution at which the studies were conducted and ethical approval was obtained from Sylhet Diabetic Hospital, Sylhet Bangladesh. Ref: S.D.A/88

Informed Consent: Informed consent was obtained from all individual participants included in the study.

References

1. The 6 Different Types of Diabetes: (5 Mar 2018). The diabetic journey. https://thediabeticjourney.com/the-6-different-types-of-diabetes

2. Statistics About Diabetes: American Diabetes Association, 22 Mar 2018. https://www.diabetes.org
3. Diabetes, World Health Organization (WHO): 30 Oct 2018. https://www.who.int/news-room/fact-sheets/detail/diabetes
4. Failure to detect type 2 diabetes early costing $700 million per year, Diabetes Australia, 8 July 2018. https://www.diabetesaustralia.com.au
5. Harris, M.I., et al.: Onset of NIDDM occurs at least 4–7 yr before clinical diagnosis. Diabetes Care **15**(7), 815–819 (1992)
6. Akter, S., et al.: Prevalence of diabetes and prediabetes and their risk factors among Bangladeshi adults: a nationwide survey. Bull. World Health Organ. **92**, 204–213A (2014)
7. Ramachandran, A.: Know the signs and symptoms of diabetes. Indian J. Med. Res. **140**(5), 579 (2014)
8. Kumar, V., Valide, L.: A data mining approach for prediction and treatment of diabetes disease. Int. J. Sci. Invent. Today (2014). ISSN 2319-5436
9. Agrawal, P., Dewangan, A.: A brief survey on the techniques used for the diagnosis of diabetes-mellitus. Int. Res. J. Eng. Technol. (IRJET). **02**(03) (2015). e-ISSN: 2395-0056; p-ISSN: 2395-0072
10. Joshi, T.N. Chawan, P.M.: Diabetes prediction using machine learning techniques. Dewangan, S. et.al. Int. J. Eng. Res. Appl. (Part -II) **8**(1), 09–13 (2018). ISSN: 2248-9622
11. Sapon, M.A., Ismail, K., Zainudin, S.: Prediction of diabetes by using artificial neural network. In: 2011 International Conference on Circuits, System and Simulation IPCSIT, vol. 7. IACSIT Press, Singapore (2011)
12. Asir, A.G., Singh, E.J., Leavline, Baig, B.S.: Diabetes prediction using medical data. J. Comput. Intell. Bioinform. **10**(1), 1–8 (2017)
13. Ahmed: Developing a predicted model for diabetes type 2 treatment plans by using data mining (2016b)
14. Rabinal, Er. Anshu Chopra2: Diabetes prediction by supervised and unsupervised learning with feature selection, **2**(5). ISSN: 2454-132
15. Mishra, V., Samuel, C., Sharma, S.K.: Use of machine learning to predict the onset of diabetes. Int. J. Recent Adv. Mech. Eng. (IJMECH) **4**(2) (2015)

Medical Diagnosis Under Uncertain Environment Through Bipolar-Valued Fuzzy Sets

Palash Dutta and Dhanesh Doley

Abstract In medical field, the necessary data or information provided by patients to medical experts are more often uncertain or vague. Fuzzy set theory is explored to deal with these types of uncertain environment and to analyze the data explained by the patients. Numerous studies have been done in the field of medicine using fuzzy sets, in particular, medical diagnosis is a prominent one. However, most of the studies leads to counterintuitive results. Therefore, in this paper, a medical diagnosis has been made using the composition of bipolar-valued fuzzy sets. It is observed that the present approach has the ability to address the medical diagnosis problem so effectively, which tallies with analytical results along with human intuitions.

Keywords Bipolar-valued fuzzy sets · Intuitionistic fuzzy sets (IFS) · Positive membership function · Negative membership function · Medical diagnosis

1 Introduction

Uncertainty is an integral part of real world problems such as engineering, medicine, environmental science, social science, etc. and it may arise due to lack of information, imprecision, vagueness etc. To evaluate such types of problems, some mathematical modeling can be used like fuzzy set theory, probability theory, etc., and each of them has some limitations. Fuzzy set was introduced by Zadeh [1], which deals with an uncertain environment. Various types of fuzzy sets are used to solve the uncertain problems such as triangular fuzzy sets, trapezoidal fuzzy sets, parabolic fuzzy sets, intuitionistic fuzzy sets, fuzzy soft sets, etc. Bipolar-valued fuzzy set is also one of them. Lees [2] initiated an extension of fuzzy set and named bipolar-valued fuzzy set in 2000. The membership degree of bipolar-valued fuzzy set ranges from [0, 1] to

P. Dutta (✉) · D. Doley
Dibrugarh University, Dibrugarh 786004, India
e-mail: palash.dtt@gmail.com

D. Doley
e-mail: dhaneshdoley01@gmail.com

M. Gupta et al. (eds.), *Computer Vision and Machine Intelligence in Medical Image Analysis*, Advances in Intelligent Systems and Computing 992,
https://doi.org/10.1007/978-981-13-8798-2_13

$[-1, 1]$, In a bipolar-valued fuzzy set, the membership degree 0 denotes that elements are irrelevant to the corresponding property. Also, the membership degree on (0, 1] denotes that elements somewhat satisfy the property and the membership degree on $[-1, 0)$ denotes that elements somewhat satisfy the suggested property.

In the literature, many authors worked on medical diagnosis under the light of fuzzy sets such as De et al. [3], discussed about medical diagnosis using IFSs. Szmidt and Kacpryzyk [4] presented a medical diagnostic reasoning using similarity measure of IFSs. Maheswari and Srivastava [5] discussed the application of IFSs on medical diagnosis. Ngana et al. [6], Dutta and Goala [7], Dutta and Talukdar [8], Vlachos and Sergiadis [9]; Own [10], Ye [11] Boran and Akay [12], Davvaz and Sadrabadi [13] also carried out medical investigation for the same IFS data. On the other hand, some applications of bipolar fuzzy sets are encountered in literature such as Mahmood et al. [14] worked on bipolar fuzzy subgroup. Da Silva Neves et al. [15] discussed decision-making problem using bipolar fuzzy set. Bipolar-valued fuzzy BCK/BCI-algebra is introduced by Said [16]. Abdullah, Aslam et al. [17] discussed decision-making problem using bipolar fuzzy soft sets. Chen and Li et al. [18] discussed an extension of bipolar fuzzy sets, i.e., m-polar fuzzy sets. Also, Mahmood, Abdullah et al. [19] discussed multi-criteria decision-making problem of bipolar-valued fuzzy sets. However, no effort has been observed in medical diagnosis using bipolar fuzzy sets.

In this paper, an attempt has been made to convert IFS medical data into bipolar form first, and then the composition of bipolar-valued fuzzy set has been performed. Later, a mathematical formulation has been adopted to defuzzify bipolar data to carry out medical diagnosis.

2 Preliminaries

In this section, we discuss about bipolar-valued and intuitionistic fuzzy sets.

2.1 Bipolar-Valued Fuzzy Set

Definition 1 Let A be any set in a universal discourse X. Then A is said to be bipolar-valued fuzzy set [2] if $A = \{(x, \mu_A^+(x), \mu_A-(x)) : x \in X\}$ where $\mu_A^+ : X \to [0, 1]$ and $\mu_A^- : X \to [-1, 0]$. That is, the positive membership degree $\mu_A^+(x)$ denotes the satisfaction degree of element x to the property corresponding to a bipolar-valued fuzzy set A and the negative membership degree $\mu_A^-(x)$ denotes the satisfaction degree of x to some implicit counter property corresponding to the bipolar-valued fuzzy set A. $\mu_A^+(x) \neq 0$ and $\mu_A^-(x) = 0$ indicates only positive satisfaction for A while $\mu_A^+ = 0$ and $\mu_A^- \neq 0$ indicates counter satisfaction property of A. It is possible for elements x to be $\mu_A^+(x) \neq 0$ and $\mu_A^-(x) \neq 0$, when the membership function of the property overlaps that of its counter property over some portion of the domain.

The reduced form of bipolar-valued fuzzy set A on the domain X is given by [20]

$$A = \{(x, \mu_A^R(x)) : x \in X\}$$

where

$$\mu_A^R : X \to [-1, 1]$$

Also, the membership degree $\mu_A^R(x)$ for the reduced representation can be derived from the above canonical form as

$$\mu_A^R(x) = \begin{cases} \mu_A^+(x), & \mu_A^- = 0 \\ \mu_A^-(x), & \mu_A^+(x) = 0; \\ f(\mu_A^+(x), \mu_A^-(x)), & \text{otherwise.} \end{cases}$$

Here, $f(\mu_A^+(x), \mu_A^-(x))$ is an aggregation function to merge a pair of positive and negative membership values into a value. Also, we can define the aggregation function $f(\mu_A^+(x), \mu_A^-(x))$ in various ways which may depend on the application domain.

Set Operation on Bipolar-Valued Fuzzy Sets: Let A_1 and A_2 be any two bipolar-valued fuzzy sets in X whose canonical representations are given by

$$\begin{aligned} A_1 =& \{(x; (\mu_{A_1}^+(x), \mu_{A_1}^-(x))) : x \in X\} \ and \\ A_2 =& \{(x; (\mu_{A_2}^+(x), \mu_{A_2}^-(x))) : x \in X\} \end{aligned}$$

respectively. Then the set operations for bipolar-valued fuzzy sets are defined as follows [20]:

$$\begin{aligned} A_1 \cup A_2 =& \{(x, \mu_{(A_1 \cup A_2)(x))} : x \in X\} \\ & \mu_{A_1 \cup A_2}(x) = (\mu_{A_1 \cup A_2}^+(x), \mu_{A_1 \cup A_2}^-(x)) \\ & \mu_{A_1 \cup A_2}^+(x) = max\{\mu_{A_1}^+(x), \mu_{A_2}^+(x)\} \\ & \mu_{A_1 \cup A_2}^-(x) = min\{\mu_{A_1}^-(x), \mu_{A_2}^-(x)\}. \\ A_1 \cap A_2 =& \{(x, \mu_{(A_1 \cap A_2)}(x)) : x \in X\} \\ & \mu_{A_1 \cap A_2}(x) = (\mu_{A_1 \cap A_2}^+(x), \mu_{A_1 \cap A_2}^-(x)) \\ & \mu_{A_1 \cap A_2}^+(x) = min\{\mu_{A_1}^+(x), \mu_{A_2}^+(x)\} \\ & \mu_{A_1 \cap A_2}^-(x) = max\{\mu_{A_1}^-(x), \mu_{A_2}^-(x)\}. \\ A_1^c =& \{(x, \mu_{A_1^c}(x)) : x \in X\} \\ & \mu_{A_1^c}(x) = (\mu_{A_1^c}^+(x), \mu_{A_1^c}^-(x)) \\ & \mu_{A_1^c}^+(x) = 1 - \mu_{A_1}^+(x) \\ & \mu_{A_1{}^c}^- = -1 - \mu_{A_1}^-(x) \end{aligned}$$

2.2 Intutionistic Fuzzy Set

Definition 2 The intuitionistic fuzzy set theory is an extension of the fuzzy set theory by Atanassov [19]. Let A be a set in the universal set X. Then A is said to be intuitionistic fuzzy set and is defined as

$$A = \{(x, \mu_A(x), \nu_A(x)) : x \in X\}$$

where the functions $\mu_A(x) : X \to [0, 1]$ defines the degree of membership of the element $x \in X$ to the set A and $\nu_A(x) : X \to [0, 1]$ defines the degree of nonmembership of the element $x \in X$ to the set A. Also, for every $x \in X$,

$$0 \leq \mu_A(x) + \nu_A(x) \leq 1$$

The amount $\pi_A(x) = 1 - (\mu_A(x) + \nu_A(x))$ is called the hesitation part or intuitionistic index, which may cater either membership degree or nonmembership degree.

Set Operation on Intuitionistic Fuzzy Sets: Let A_1 and A_2 be any two intuitionistic fuzzy sets in X which are defined by

$$A_1 = \{(x, \mu_{A_1}(x), \nu_{A_2}(x)) : x \in X\}$$

and

$$A_2 = \{(x, \mu_{A_2}(x), \nu_{A_2}(x)) : x \in X\},$$

respectively, and where $\mu_{A_1}(x)$ and $\nu_{A_1}(x)$ are, respectively, the membership degree and nonmembership degree of x in A_1. Similarly, for A_2 also.
The set operations of A_1 and A_2 are defined as follows [20]:

$$\begin{aligned}
A_1 \cup A_2 =& \{(x, \mu_{A_1 \cup A_2}(x), \nu_{A_1 \cup A_2}(x)) : x \in X\} \\
& \mu_{A_1 \cup A_2}(x) = max\{\mu_{A_1}(x), \mu_{A_2}(x)\} \\
& \nu_{A_1 \cup A_2}(x) = min\{\nu_{A_1}(x), \nu_{A_2}(x).\} \\
A_1 \cap A_2 =& \{(x, \mu_{A_1 \cap A_2}(x), \nu_{A_1 \cap A_2}(x)) : x \in X\} \\
& \mu_{A_1 \cap A_2}(x) = min\{\mu_{A_1}(x), \mu_{A_2}(x)\} \\
& \nu_{A_1 \cap A_2}(x) = max\{\nu_{A_1}(x), \nu_{A_2}(x).\} \\
A_1^c =& \{(x, \mu_{A_1^c}(x), \nu_{A_1^c}(x)) : x \in X\} \\
& \mu_{A_1^c}(x) = \nu_{A_1}(x) \\
& \nu_{A_1^c}(x) = \mu_{A_1}(x).
\end{aligned}$$

3 Relationship Between IFSs and Bipolar-Valued Fuzzy Sets

The canonical representation of a bipolar-valued fuzzy set A is $A = \{(x, (\mu_A^+(x), \mu_A^-(x))) : x \in X\}$ and representation of intuitionistic fuzzy set (IFS) [20] A is $A = \{(x, \mu_A(x), \nu_A(x)) : x \in X\}$. In comparison, we get $\mu_A^+(x) = \mu_A(x)$ and $\mu_A^-(x) = -\nu_A(x)$, which look like similar to each other. However, they are different from each other in some circumstances such as in bipolar-valued fuzzy sets, the positive membership degree $\mu_A^+(x)$ characterizes the extent that the element x satisfies the property A and the negative membership degree $\mu_A^-(x)$ characterizes the extent that the element x satisfies an implicit counter property of A. But in case of intuitionistic fuzzy sets, the membership degree $\mu_A(x)$ denotes the degree that the element x satisfies the property A and the membership degree $\nu_A(x)$ indicates the degree that the element x satisfies the not-property of A. As a counter property is not usually equivalent to not-property, both bipolar-valued fuzzy sets and intuitionistic fuzzy sets are the different extensions of fuzzy sets.

4 Medical Diagnosis

This section presents an application of bipolar-valued fuzzy set in the medical field. It is seen that medical diagnosis problems are dealing with uncertainty, imprecision, vagueness, etc. To explore this types of problems, bipolar-valued fuzzy set is introduced. Here, to perform medical diagnosis, **max–min** composition of bipolar-valued fuzzy set has been introduced.

Suppose that, $P = \{p_i\}$ be a set of patients, $S = \{s_i\}$ be a set of symptoms and $D = \{d_i\}$ ba set of diseases (i=1,2,3, …). The medical diagnosis will be performed with mainly three steps:

Step1: Determination of patient symptoms.

Step2: Formulation of medical knowledge based on bipolar-valued fuzzy relations.

Step3: Determination of diseases on the basis of composition of bipolar-valued fuzzy relations.

Let $R_2(S \to D)$ be a bipolar-valued fuzzy relation and also construct a bipolar-valued fuzzy relation R_1 from the set patients(P) to the set of symptoms(S). Then the composition R_3 of bipolar-valued fuzzy relations R_2 and R_1, i.e., $R_3 = R_2 \circ R_1$ describes the state of patients p_i in terms of the disease as a bipolar-valued fuzzy relation from the set of patient(P) to the set of diseases(D) with positive membership function is given by

$$\mu_{R_3}^+(p_i, d) = max[min\{\mu_{R_1}^+(p_i, s), \mu_{R_2}^+(s, d)\}]$$

and the negative membership function is given by

$$\mu^-_{R_3}(p_i, d) = min[max\{\mu^-_{R_1}(p_i, s), \mu^-_{R_2}(s, d)\}]$$

$\forall p_i \in P$ and $d \in D$.
For ranking of bipolar-valued fuzzy sets, we adopt the following mathematical formula [21]:

$$Rank = \frac{\mu^+_{R_3}(p_i, d) + (1 - \mu^-_{R_3}(p_i, d))}{2}.$$

4.1 Drawbacks of Existing Approaches

It is well known that uncertainty is an integral component of our real-life problem. Specially, medical diagnosis problems deals with an uncertain environment. Fuzzy set theory plays an important role for such kinds of problems, which are associated with uncertainty. Many researchers worked on medical diagnosis using different types of fuzzy sets, such as IFSs, PFSs, IVFSs, etc. In 2001, De et al. [3] first presented medical diagnosis using IFSs. Then, Szmidt and Kacprzyk [22] has criticized De et al. [3] approach and performed another study in medical diagnosis using the same IFSs medical data, but the results were obtained to be different. Several other studies have been done using the same IFSs medical data such as Maheswari and Srivastava [5], Ngana et al. [6], Dutta and Goala [7], Dutta and Talukdar [8], Vlachos and Sergiadis [9], Own [10], Ye [11] Boran and Akay [12], Davvaz and Sadrabadi [13], however, different chaotic and counterintuitive results were obtained. But, the present study wants to establish that the results obtained by this method is logical, analytically correct, and represent human intuitions.

4.2 Case Study

To perform medical diagnosis, consider four patients Rahul, Ratul, Rakesh, and Rabin which are admitted in a hospital in Jorhat, Assam. The symptoms expressed by patients to the medical expert are temperature, headache, stomach pain, cough, and chest pain. Consequently, the set of patients

$P = \{Rahul, Ratul, Rakesh, Rabin\}$ and the set symptoms

$S = \{Temperature, Headache, Stomach - pain, Cough, Chest - pain\}$.

Let $D = \{ViralFever, Malaria, Typhoid, Stomach - problem, Heart - probleme\}$ be the set of diseases given by the expert corresponding to the above symptoms S.

The bipolar-valued fuzzy relation $R_1(P \to S)$ is present in Table 1 and the bipolar-valued fuzzy relation $R_2(S \to D)$ is presented in Table 2. Also, the composition relation $R_3 = R_2 \circ R_1$ is presented in Table 3.

Table 1 Patient–symptoms relation

R_1	Temperature	Headache	Stomach-pain	Cough	Chest pain
Rahul	(0.8, −0.1)	(0.6, −0.1)	(0.2, −0.8)	(0.6, −0.1)	(0.1, −0.6)
Ratul	(0, −0.8)	(0.4, −0.4)	(0.6, −0.1)	(0.1, −0.7)	(0.1, −0.8)
Rakesh	(0.8, −0.1)	(0.8, −0.1)	(0, −0.6)	(0.2, −0.7)	(0, −0.5)
Rabin	(0.6, −0.1)	(0.5, −0.4)	(0.3, −0.4)	(0.7, −0.2)	(0.3, −0.4)

Table 2 Symptoms–disease relation

R_2	Viral fever	Malaria	Typhoid	Stomach problem	Chest problem
Temperature	(0.4, 0.0)	(0.7, 0.0)	(0.3, −0.3)	(0.1, −0.7)	(0.1, −0.8)
Headache	(0.3, −0.5)	(0.2, −0.6)	(0.6, −0.1)	(0.2, −0.4)	(0.0, −0.8)
Stomach-pain	(0.1, −0.7)	(0.0, −0.9)	(0.2, −0.7)	(0.8, 0.0)	(0.2, −0.8)
Cough	(0.4, −0.3)	(0.7, 0.0)	(0.2, −0.6)	(0.2, −0.7)	(0.2, −0.8)
Chest pain	(0.1, −0.7)	(0.1, −0.8)	(0.1, −0.9)	(0.2, −0.7)	(0.8, −0.1)

Table 3 Patient–disease relation

R_3	Viral fever	Malaria	Typhoid	Stomach problem	Chest problem
Rahul	(0.4, −0.7)	(0.7, −0.8)	(0.6, −0.7)	(0.2, −0.6)	(0.2, −0.8)
Ratul	(0.3, −0.7)	(0.2, −0.8)	(0.4, −0.8)	(0.6, −0.7)	(0.1, −0.8)
Rakesh	(0.4, −0.6)	(0.7, −0.6)	(0.6, −0.6)	(0.2, −0.7)	(0.2, −0.7)
Rabin	(0.4, −0.4)	(0.7, −0.4)	(0.5, −0.4)	(0.3, −0.4)	(0.3, −0.4)

Table 4 Ranking value of patient–disease relation

R_3	Viral fever	Malaria	Typhoid	Stomach problem	Chest problem
Rahul	1.05	**1.25**	1.15	0.9	1.0
Ratul	1.0	1.0	1.1	**1.15**	0.95
Rakesh	1.0	**1.15**	1.1	0.95	0.95
Rabin	0.9	**1.05**	0.95	0.85	0.85

Finally, the calculated rank value using by the above ranking formula, i.e.,

$$Rank = \frac{\mu^{+}_{R_3}(p_i, d) + (1 - \mu^{-}_{R_3}(p_i, d))}{2}$$

is described in Table 4.

From the above Table 4, it is observed that the maximum numerical value arises in the first row is **1.15** which is associated with the disease *Malaria*. Therefore, it is cleared that *Rahul* is likely suffering from *Malaria*. Similarly, it can be concluded that *Ratul*, *Rakesh*, and *Rabin* is suffering from the disease *Stomachproblem*, *Malaria*, and *Malaria* respectively.

5 Conclusion

In this paper, we studied bipolar fuzzy sets and its comparison with IFS. Here, we observed that bipolar fuzzy set has a relationship with IFS. Also, here, a case study on medical diagnosis has been made through bipolar-valued fuzzy sets and it is observed that a patient is suffering from a disease if he has the maximum ranking value associated with that disease. Furthermore, the present approach provides logically and analytically true results which also have the ability to human intuitions.

References

1. Zadeh, L.A.: Fuzzy sets. Inf. Control **8**, 338–356 (1965). https://doi.org/10.1016/S0019-9958(65)90241-X
2. Lee, K.M.: Bipolar-valued fuzzy sets and their operations. In: Proceedings of the Conference on Intelligent Technologies, pp. 307–312, Bangkok, Thailand (2000)
3. De, S.K., Biswas, R., Roy, A.R.: An application of intuitionistic fuzzy sets in medical diagnosis. Fuzzy Sets Syst. **117**(2), 209–213 (2001)
4. Szmidt, E., Kacprzyk, J.: A similarity measure for intuitionistic fuzzy sets and its application in supporting medical diagnostic reasoning. In: International Conference on Artificial Intelligence and Soft Computing. Springer, Berlin, Heidelberg (2004)
5. Maheshwari, S., Srivastava, A.: Study on divergence measures for intuitonistic fuzzy sets and its application in medical diagnosis. J. Appl. Anal. Comput. **6**(3), 772–789 (2016)
6. Ngana, R.T., Sonb, L.H., Cuongc, B.C., Alid, M.: H-max distance measure of intuitionistic fuzzy sets in decision making. Appl. Soft Comput.
7. Dutta, P., Goala, S.: Fuzzy decision making in medical diagnosis using an advanced distance measure on intuitionistic fuzzy sets. Open Cybern. Syst. J. **12**(1) (2018)
8. Talukdar, P., Dutta, P.: Disease diagnosis using an advanced distance measure for intuitionistic fuzzy sets. Int. Res. J. Microbiol. **7**(2), 029–042 (2018)
9. Vlachos, I.K., Sergiadis, G.D.: Intuitionistic fuzzy information applications to pattern recognition. Pattern Recognit. Lett. **28**(2), 197–206 (2007)
10. Own, C.M.: Switching between type-2 fuzzy sets and intuitionistic fuzzy sets: an application in medical diagnosis. Appl. Intel. **31**, 283–291 (2009)
11. Ye, J.: Cosine similarity measures for intuitionistic fuzzy sets and their applications. Math. Comp. Model. **53**(1), 91–7 (2011)
12. Boran, F.E., Akay, D.: A biparametric similarity measure on intuitionistic fuzzy sets with applications to pattern recognition. Inf. Sci. **10**(255) , 45–57 (2014)
13. Davvaz, B. , Hassani, S.E.: An application of intuitionistic fuzzy sets in medicine. Int. J. of Biomath. **9**, 1650037 (2016)
14. Mahmood, T., Munir, M.: On bipolar fuzzy subgroups. World Appl. Sci. J. **27**(12), 1806–1811 (2013)

15. Neves, R.D.S., Livet, P.: Bipolarity in human reasoning and affective decision making. Int. J. Intel. Syst. **23**(8), 898–922 (2008)
16. Saeid, A.B.: Bipolar-valued fuzzy BCK/BCI-algebras. World Appl. Sci. J. **7**(11), 1404–1411 (2009)
17. Abdullah, S., Aslam, M., Ullah, K.: Bipolar fuzzy soft sets and its applications in decision making problem. J. Intell. Fuzzy Syst. **27**(2), 729–742 (2014)
18. Chen, J., Li, S., Ma, S., Wang, X.: Polar fuzzy sets: an extension of bipolar fuzzy sets. Sci. World J. (2014)
19. Mahmood, T., Abdullah, S., Bilal, M., Rashid, S.: Multiple criteria decision making based on bipolar valued fuzzy sets. In: Atanassov, K.T (ed) Intuitionistic Fuzzy sets. Fuzzy Sets and Systems, vol. 20, no. 1, pp. 87–96 (1986). https://doi.org/10.1016/S0165-0114(86)80034-3
20. Lee, K.M., Lee, K.M., Cios, K.J.: Comparison of interval-valued fuzzy sets, intuitionistic fuzzy sets, and bipolar-valued fuzzy sets. In: Computing and Information Technologies: Exploring Emerging Technologies, pp. 433–439 (2001)
21. Tamani, N.: A bipolar approach for intuitionistic fuzzy alternative ranking. In: IEEE International Conference on Fuzzy Systems (FUZZ-IEEE), pp. 1–8. IEEE (2015)
22. Szmidt, E., Kacprzyk, J.: Intuitionistic Fuzzy Sets in Some Medical Applications. Fuzzy Days (2001)

Design and Analysis of Novel Room Temperature T-Ray Source for Biomedical Imaging: Application in Full Body Prosthetics

Saikat Adhikari, Singam Jayanthu and Moumita Mukherjee

Abstract The authors have designed and studied the superlattice terahertz device for accurate detection of cancer cell in a Full Body Prosthetic (FBP). For this, a generalized nonlinear simulator is developed and the same is verified by comparing the results with those of experimental observation. The model predicts that identification of cancerous cell in FBP could be done satisfactorily by analyzing corresponding thermographs. For T-Ray source and detector, the authors have considered p++ -n- - n - n++ type Mixed Tunneling Avalanche Transit Time (MITATT) Device at 0.1 THz. The study reveals that the proposed device is capable of developing 10 W level of fundamental harmonic power at around 100 GHz. The authors have also studied the effects of modulation on electric field profiles for different phase angles. The simulator incorporates the physical and electrical properties of GaN/AlN superlattice, which include temperature and field dependent carrier ionization rates, saturation velocity of charge carriers, mobility, inter-subband tunneling and drift velocity overshoot effects as well as hot carrier effects inter-band scattering of electron-hole pairs in superlattice region. An equivalent circuit model is developed and analyzed for obtaining impedance and admittance characteristics. To the best of authors' knowledge this is the first report on large-signal modeling of THz Solid State imaging unit for thermographic analysis of malignant tumors in Full Body Prosthetics (FBP).

Keywords FBP · Solid state T-Ray source · Nonlinear large-signal analysis · GaN/AlN superlattice · Room temperature T-Ray radiation system · Radiation thermographs

---

S. Adhikari · M. Mukherjee (✉)
Department of Physics, Adamas University, Kolkata 700126, India
e-mail: mm.adamasuniv@hgmail.com

S. Adhikari
e-mail: spgrar@live.com

S. Jayanthu
NIT – Rourkella, Rourkela, India

M. Gupta et al. (eds.), *Computer Vision and Machine Intelligence in Medical Image Analysis*, Advances in Intelligent Systems and Computing 992,
https://doi.org/10.1007/978-981-13-8798-2_14

1 Introduction

The terahertz region (1 THz = 1012 Hz) lies in between the microwave and infrared regime of electromagnetic spectrum. This has now become a promising area of R&D activities in the diversified field of Physics, Chemistry, Engineering, and Medical and Biological Sciences. The unique property of T-Ray is its low photon energy which intern is beneficial for medical applications owing to its nonionizing nature. The presence of malignancy in human blood cell causes increase in tissue water content. This acts as a contrast in T-Ray imaging. In spite of its huge application possibilities, the Biomedical Instrument Industries are still lagging in full utilization of this range of EM Spectrum for the noninvasive and nonhazardous imaging of human body. This is because of the lack of suitable room temperature and compact sources and detectors. Most of the available T-Ray sources are bulky and low temperature and therefore not suitable for Biomedical applications. Considering the ever growing need of the development of T-Ray sources for biomedical purpose, the authors have proposed a new class of solid-state room temperature device that can be used as a potential T-Ray source for the identification of malignant tumors in organ placed in a Full Body Prosthetic (FBP) arrangement.

A full body prosthetic (FBP) is an artificial system holding the life support system for an isolated brain or transplanted head. Due to the experimental nature of artificial organ technologies, an artificial body could be designed to house biological organs from a donor. A noninvasive and safe method of biomedical scanning would be essential to reduce the risk of infection and structural weakness if the FBP's internal system is exposed. The terahertz band lies between the microwave and infrared regions of the electromagnetic spectrum and the radiation has very low photon energy, and thus it does not pose any ionization hazard for biological tissues. The radiation has unique absorption spectra due to intermolecular vibrations in this region that have been found in different biological materials. This is the uniqueness of T-Ray-imaging in medical applications for providing complimentary information to the existing imaging techniques. For checking the viability of T-Ray imaging for an FBP, a full-scale simulator will be developed. The Comsol based model will consider the healthy and malignant cells under various operating conditions inside an FBP, so as to approximate a human body. The simulated Terahertz imaging device will be used to locate the condition and location of the target cells. In addition to being relevant to detecting malignant tumors in a modeled FBP, the findings could further be useful in analysis of customized bioreactors for the biotechnology and pharmaceutical industries.

The present paper will report (i) Design and Characterization of an exotic Avalanche Transit Time (ATT) device for T-Ray generation, (ii) Design of a suitable T-Ray radiation system; (iii) T-Ray imaging snap shots of malignant tumor located in a specially designed FBP. The authors have designed a ATT device where carrier generation is contributed by both Avalanche multiplication and inter-band tunneling phenomenon. The resultant device will operated in MITATT (Mixed Impact-ionization Tunneling ATT) mode and corresponding power generation will be in

THz frequency regime. The proposed structure is a Hexagonal Wz-GaN/AlN superlattice of periodicity four with asymmetrical doping and width distribution in the active region of the device. Superlattice structure with asymmetrical doping profile in central active region results in spatial separation of mobile electrons and holes within the central region of the designed ATT device. This improves the electrical/electronic/transport properties such as carrier lifetime, saturation velocity, and mobility significantly. The band to band transition and drifting of charge carriers in the active region of the device further induces a current-pulse in the external-circuit and this generates an oscillation of desired frequency in THz region. The authors have earlier developed and published a nonlinear, self-consistent, large-signal simulator for the realistic modeling and analysis of MITATT devices [1, 2]. In the present study, the authors have used that simulator with some important modification for the various quantum aspect of carrier transition in asymmetrically doped GaN/AlGaN superlattice structure. Mixed Quantum Classical Drift-Diffusion model is used for solving nonlinear Poisson and charge continuity equations subject to appropriate boundary conditions [3]. The validity of the model is established by comparing the simulated data with experimental findings.

A huge research work has been done, by this time, with Wide Band Gap (WBG) semiconductor-based MITATT/IMPATT devices in current years [1, 2]. The research works are mostly focused on IMPATT mode of operation with flat type doping density distribution. The published studies have established the potentiality/superiority of wide-bandgap SiC, III–V GaN, and Si/SiC materials for generating T-Ray signal with a medium to low efficiency [1, 2]. To the best of authors' knowledge asymmetrically doped superlattice MITATT devices are not presently available in published literature. High-frequency oscillation generation requires high mobility of charge carrier in transit. Specially designed superlattice structure is much promising from this aspect of study; this has prompted the authors to choose such exotic doping profiles for designing of a room temperature and efficient power source at Terahertz frequency region.

Wide band gap materials (III–V and IV–IV compound semiconductors) are promising for developing high power efficient ATT devices. Power output from an ATT device depends upon the saturation velocity and critical electric field at breakdown of the base semiconductor. GaN and AlN, having saturation velocity ~2 × 105 m/s, critical breakdown field ~2 × 108 V/m, are expected to be a potential pair for developing a superlattice structure. The inherent mobility of AlN/GaN reduces transit time of carriers through the active region of the device. This makes the device suitable for oscillating at THz (0.1 THz to 10 THz) frequency region. Moreover, the lattice mismatch in between sapphire substrate and epilayer AlN/GaN is minimum compared to flat GaN epilayer [4]. Thus the authors have chosen AlN/GaN superlattice for designing the high power, high-frequency ATT device.

Worldwide physicians are concerned for the early diagnosis of cancer in human body so as to ensure that patient's life could be saved. Most of the noninvasive imaging techniques, those are commonly available nowadays, mostly rely upon X-Ray. However, X-Ray is an ionizing radiation and it's a secondary cause of malignancy. Moreover, early diagnosis of malignant tumor is the biggest unsolved issue/problem as X-Ray can only detect tumor of dimension >7 mm. T-Ray, on the other hand, is non-ionizing and thus expected to identify cancerous tumor of less than 1 mm diameter. This possibility is thoroughly studied by the authors in the present paper by designing a computer-based FBP system with malignant and nonmalignant cell/tissues tumors of various dimensions within the designed FBP. COMSOL multiphysics based semi-conductor/RF module and heat transfer module are used for this purpose.

2 Methodology

This part of the research article will deal with the design and simulation methodology, imposed boundary conditions and the device dimension details. The work flow diagram is shown in Fig. 1.

2.1 *Quantum Modified Non-linear Drift-Diffusion (QMNLDD) Model for ATT Devices*

The asymmetrically doped AlN/GaN-ATT (p++ -n- - n+ - n++ doping profile) Terahertz source and detector have been designed and analyzed in the paper. The physical properties including thermal/electrical/electronic properties of AlN/GaN materials along the symmetric axis of the device are shown in Table 1. The authors have made a generalized, non-linear Large-signal (L-S) simulation to get original / realistic view of the device characteristics under various operating conditions. For each instant of time, the physical properties, including, electric field, charge carrier current components and recombination current are obtained by solving the non-linear

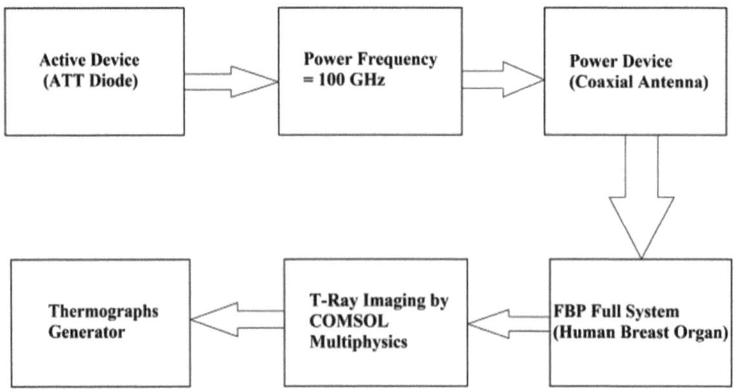

Fig. 1 Work flow diagram of the T-Ray scanning and imaging system developed in-situ

Table 1 Material parameters of different semiconductors

S. No	Attribute	Symbol with unit	Si	GaAs	6H-SiC	4H-SiC	GaN	AIN
1	Bandgap	E_g (Electron Volt)	1.12	1.43	3.03	3.26	3.45	6.05
2	Dielectric Constant	ε_r	11.9	13.1	9.66	10.1	9.00	9.14
3	Electric Breakdown Field	E_c (kV/cm)	300	400	2,500	2,200	2,000	
4	Electron Mobility	μ_n (cm^2/V s)	1500	8500	500 80	1000	1250	300
5	Hole Mobility	μ_p (cm^2/V s)	600	400	101	115	850	–
6	Thermal Conductivity	λ (W/cm K)	1.5	0.46	4.9	4.9	1.3	2.85
7	Saturated Electron Drift Velocity	v_{sat} ($\times 10^7$cm/s)	1	1	2	2	2.2	1.6

field and carrier transport equations, i.e. Poisson's equation and combined current continuity equations for different modulation factors at the boundaries of the active region, subject to satisfaction of appropriate boundary conditions. The authors have considered the effect of introducing a buffer n-bump layer of appropriate doping density in between the substrate and epi-layer.

$$\frac{\partial^2}{\partial x^2} V(x,t) = -\frac{q}{\varepsilon}[N_d(x,t) - N_a(x,t) + C_p(x,t) - C_n(x,t)] \tag{1}$$

$$\frac{\partial}{\partial x} p(x,t) = -(\frac{1}{q})\frac{\partial}{\partial x} J_p(x,t) + G_p(x,t) - R_p(x,t) \tag{2}$$

$$\frac{\partial}{\partial x} n(x,t) = (\frac{1}{q})\frac{\partial}{\partial x} J_n(x,t) + G_n(x,t) - R_n(x,t) \tag{3}$$

$$J_p(x,t) = -q\mu_p[C_p(x,t)\frac{\partial}{\partial x} V(x,t) + (\frac{K_B T_j}{q})\frac{d}{\partial x} C_p(x,t)] \tag{4}$$

$$J_n(x,t) = -q\mu_n[C_n(x,t)\frac{\partial}{\partial x} V(x,t) - (\frac{K_B T_j}{q})\frac{d}{\partial x} C_n(x,t)] \tag{5}$$

$$J_t(x,t) = J_n(x,t) + J_p(x,t) \tag{6}$$

where $J_{p,n}(x,t)$ denotes electron and hole current density, $V(x,t)$ is electric potential, $J_t(x,t)$ denotes the total current density, $C_{p,n}(x,t)$ is for charge carrier concentration, $G_{p,n}(x,t)$ is for carrier generation rate, $R_{p,n}(x,t)$ denotes the carrier recombination rates, $N_a(x,t)$ and $N_d(x,t)$ are the the electron and hole current densities, respectively, $\mu_{p,n}$, ε, T_j are the mobility of electrons and holes, permittivity, junction temperature respectively.

The carrier generation rates are obtained due to the avalanche phenomenon and band to band tunneling of electron and hole. It can be written as-

$$G_{p,n}(x,t) = G_{A_{p,n}}(x,t) + G_{T_{p,n}}(x,t) + G_{ph_{p,n}}(x,t) \tag{7}$$

where, $G_{A_{n,p}}(x,t)$, $G_{T_{p,n}}(x,t)$ and $G_{ph_{p,n}}(x,t)$ represent the avalanche generation rates and tunnel carrier current generation rates and opto-generation rate respectively. The avalanche carrier generation rates for electron and hole can be expressed as-

$$G_{A_p}(x,t) = G_{A_n}(x,t) = \alpha_p(x,t)v_p(x,t)C_p(x,t) = \alpha_n(x,t)v_n(x,t)C_n(x,t)$$

where, $\alpha_{p,n}$, $v_{p,n}$ are the ionization-rate and drift velocities of the charge carriers respectively. The electron tunneling generation in GaN/AlN is expressed as

$$G_{T_n}(x,t) = a_T E^2(x,t)\exp[1 - \frac{b_T}{E(x,t)}].$$

where, $E(x,t)$ represents the electric field. The coefficients a_T and b_T can be determined by

$$a_T = \frac{q}{8\pi\hbar^2}(\frac{m_n^*}{E_g})^{\frac{1}{2}}, b_T = \frac{1}{2q\hbar}(\frac{m_n^* E_g}{2})^{\frac{1}{2}}$$

where, E_g is the band gap energy introduced in AlN/GaN superlattice by means of doping, m_n^* is the effective mass of electron, $\hbar(\frac{h}{2\pi})$ is the normalized Planck's constant, q (1.6×10^{-19} C) is charge of the electron and h (6.625×10^{-34}) is the Planck's constant. The tunnel induced hole generation rate can be expressed as- $G_{T_p}(x,t) = G_{T_n}(x',t)$. The tunnel induced hole-generation rate at x is the function of electron generation rate due to tunneling at x'. where, $(x - x')$ is the spatial separation in between valance and conduction band at the same energy level. It can be obtained from the energy band diagram of p++ -n- - n+ - n++ device.

2.2 *Simulaion of FBP Model*

Comsol Multiphysics Simulator is used for designing an equivalent FBP model with cylindrical geometry. The dimension of the Cylinder is as follows:

120 mm in length and 50 mm in diameter

2.3 *Comsol Thermographic Model of T-Ray Radiation System*

Comsol Multiphysics Electromagnetic Module is used for designing T-Ray radiation system and corresponding generation of thermographs. Hyperthermic oncology and relevant models coupled with EM Modules that include bioheat equations are used for this purpose. The model takes the advantage of rotational symmetry which intern allows modeling in quasi 3D cylindrical coordinates with an appropriate selection of fine meshing to achieve excellent accuracy. The model uses frequency domain formulation. T-Ray radiation source/antenna is embedded in a FBP along its axis. Initially, the FBP is considered to be filled with nonmalignant cell and thereafter with malignant cell of appropriate permittivity and thermal conductivity values. The radiation coming out from the source has been absorbed by the surroundings cells and generates heating effects according to the electrical properties of malignant/nonmalignant cells. Due to the more water contains in malignant cell compared to its nonmalignant counter parts, thermal gradient would vary considerably and the authors have accurately studied the corresponding thermographs to detect the presence of malignant cell in FBP. In addition to heat transfer equation, the model computes cell damage integral as well. The T-Ray radiation source distribution decays gradually as a function of distance from the source. The authors have considered the electrical and thermal properties of malignant and nonmalignant cells from published literature [5].

3 Result and Discussion

The electric field snapshots are shown in Fig. 2 for different phase angles. It is depicted that the device breaks down at a critical electrical field of 1.1×10^8 V/m. The active region width ~0.8 μm. The effects of modulation at different phase angle are shown in the figures. Figure 3 depicts the admittance characteristics of the simulated T-Ray source for different operating temperatures. It is observed that the peak frequency of oscillation at 300 K is 100 GHz and the same elevated to 107 GHz for an increase of junction temperature up to 600 K. The avalanche frequency of oscillation is observed to be 48 GHz. Figure 4 shows temperature-dependent negative resistivity plots of the active device. The peak resistivity value at 300 K is found to be 5×10^{-2} Ωm. The study also reveals that the value of negative resistivity gradually decreases with increasing temperature and at 600 K the value reduces to ~30%. The profile clearly indicates that the possibility of generation of RF power is more in the mid active region. Figure 5 shows the designed cylindrical FBP. Figure 6 denotes the T-Ray Thermographs of Malignant and Nonmalignant cells in FBP. In case of normal fatty breast tissue the temperature rise, as a result of absorption of T-Ray radiation, is insignificant (almost in between 300 and 310 K), whereas the temperature variation and enhancement is quite significant in presence of malignant breast tissues. The corresponding thermographs reveals the temperature variation between 310 and 550 K. This increase of temperature is due to the presence of more water

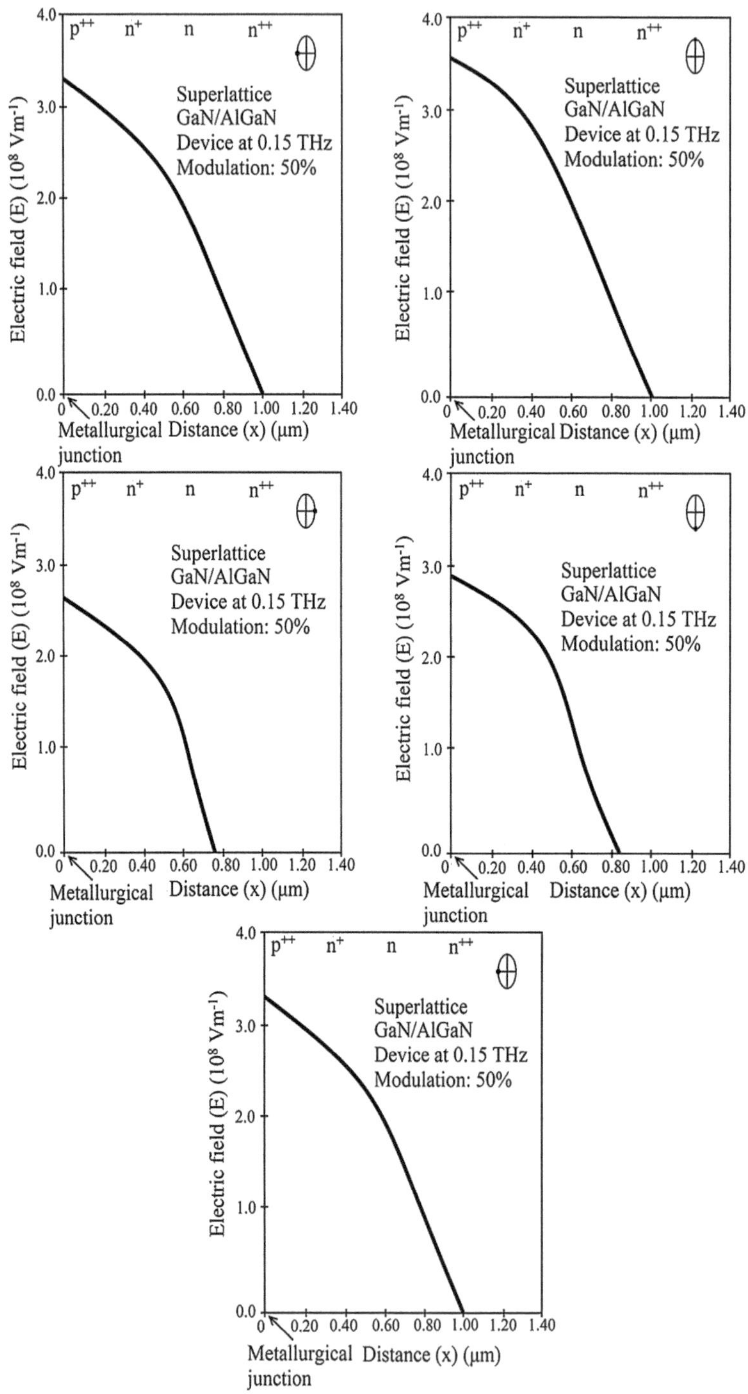

Fig. 2 Electric field properties of Wz-GaN/AIN superlattice ATT device at W-band

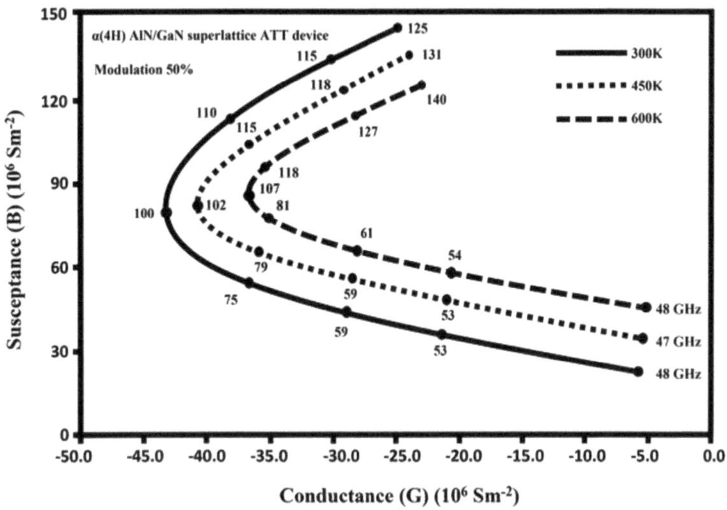

Fig. 3 Temperature-dependent admittance plots of Wz-GaN/AIN superlattice ATT device at W-band

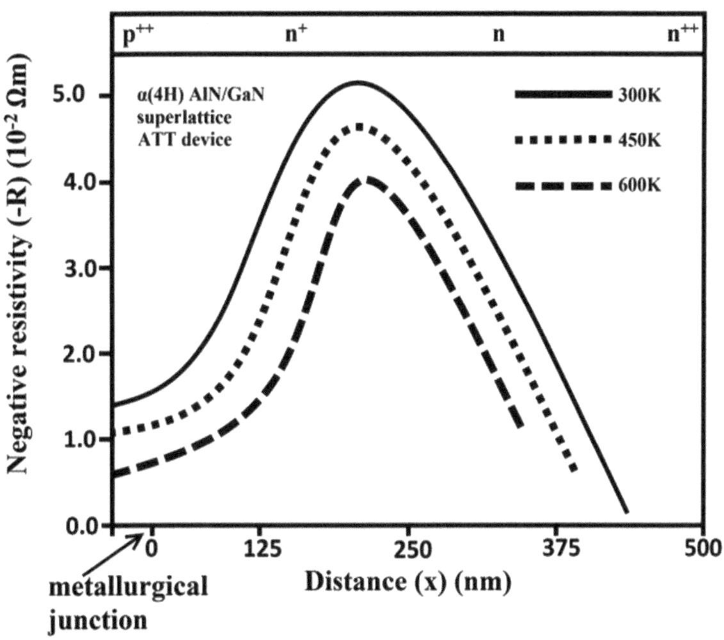

Fig. 4 Temperature-dependent negative resistivity plots of Wz-GaN/AIN superlattice ATT device at W-band

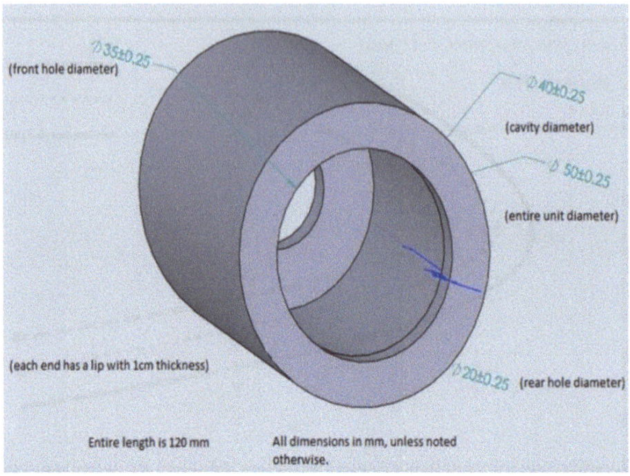

Fig. 5 Sechametic Diagram of FBP

in cancer affected cell in breast organ. The increase in temperature is more near the T-Ray radiation source and decreases gradually with distance. The dimension of the malignant tumor has been considered to be less than 1 mm. The published literature, dealt with X-Ray radiation, shows that malignant tumor of such a small dimension could not be predicted with such accuracy by simply adopting a cost-effective, room temperature, and easy technique [5].

4 Conclusion

A generalized Mixed Quantum Modified Nonlinear Drift-Diffusion (QMNLDD) simulator for designing and studying GaN/AlN exotic MITATT device has been developed by the authors. The necessity of incorporation of superlattice doping and properties in conventional model is to improve the high-frequency electronic/electrical and thermal properties of the MITATT Device. GaN/AlN asymetrical superlattice is found to be a good replacement of conventional GaN flat profile devices as far as improved admittance, electrical field profile, power output, and efficiency are concerned. T-Ray Radiation Thermographs clearly establish the accuracy level of T-Ray imaging technique in diagnosis of malignant breast tumor of <1 mm diameter. The study, for the first time, establishes the superiority of GaN/AlN superlattice based T-Ray Radiation source in hypothermic/thermal analysis of malignancy when the affected organ is inside a designed cylindrical FBP. To the best of authors' knowledge, this is the first report on asymmetrical superlattice MITATT/ATT Device in noninvasive low cost and accurate identification of Breast Cancer.

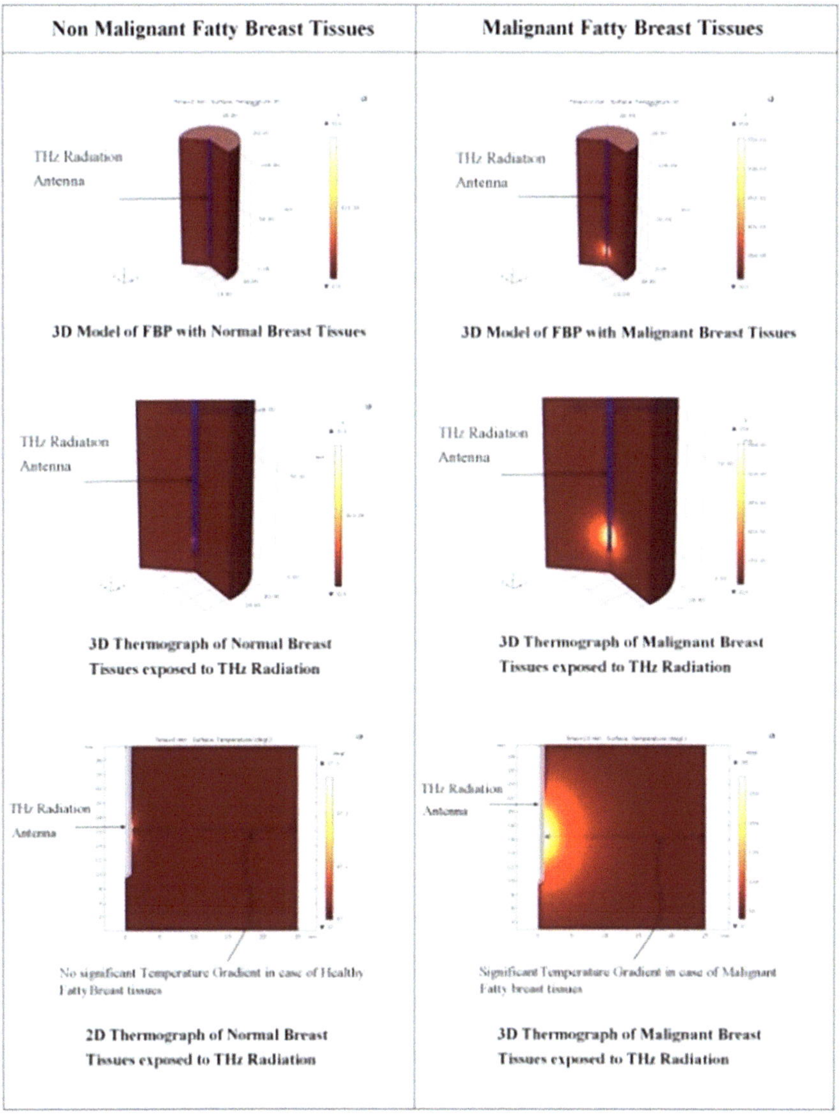

Fig. 6 Simulation thermographs: THz thermal imaging of normal and malignant human breast tissues

References

1. Mukherjee, M., Mazumdar, N., Roy, S.K.: GaN IMPATT diode: a photo-sensitive high power terahertz source. Semicond. Sci. Technol. **22**, 1258 (2007)
2. Mukherjee, M., Mazumdar, N., Roy, S.K.: Photosensitivity analysis of gallium nitride and silicon carbide terahertz IMPATT oscillators: comparison of theoretical reliability and study on experimental feasibility. IEEE Trans. Device Mater. Reliab. **8**, 608–620 (2008)
3. Kundu, A., Adhikari, S., Das, A., Ray, M., Mukherjee, M.: Design and characterisation of asymetrical superlattice Si/4H-SiC pin photo-diode array: a potential opto-sensor for future applications in biomedical domain. Microsyst. Technol. (2018). https://doi.org/10.1007/s00542-018-4119-4
4. NSM archieve, NSM Archive: Physical properties of semiconductors. www.ioffe.ru/SVA/NSM/Semicond/
5. George, T., Rufus, E., Alex, Z.C.: Simulation of microwave induced thermo-accoustical imaging technique for cancer detection. ARPN J. Eng. Appl. Sci. **10** (2015)

Author Index

M. Gupta et al. (eds.), *Computer Vision and Machine Intelligence in Medical Image Analysis*, Advances in Intelligent Systems and Computing 992,
https://doi.org/10.1007/978-981-13-8798-2

MIX
Papier aus verantwortungsvollen Quellen
Paper from responsible sources
FSC® C105338

If you have any concerns about our products,
you can contact us on
ProductSafety@springernature.com

In case Publisher is established outside the EU,
the EU authorized representative is:
Springer Nature Customer Service Center GmbH
Europaplatz 3, 69115 Heidelberg, Germany

Printed by Libri Plureos GmbH
in Hamburg, Germany